COLORFUL
MINERAL

By ANTHONY C.
TENNISSEN, Ph. D.

**Associate Professor
Lamar University**

Photographs by Werner Lieber

STERLING PUBLISHING CO., Inc. NEW YORK

Oak Tree Press Co., Ltd. London & Sydney

INTRODUCTION

Have you ever hunted for and collected sea shells, leaves and nuts, butterflies, or insects? Most of us have, and yet we are not all conchologists, botanists, lepidopterists or entomologists. We simply enjoy collecting or looking up the various specimens we find. Have you ever thought of collecting the many beautiful abundant minerals in nature? Has the task of identification seemed overwhelming? Part of the fun of finding specimens, of course, is in knowing what you are saving.

The science of mineralogy seems so complex, involving so much pre-knowledge, that it is little wonder that people sometimes shy away from it. However, you don't have to be a trained mineralogist to enjoy mineral hunting any more than you have to be a lepidopterist to enjoy butterflies. This little book is a simple basic introduction to identifying the more colorful minerals, and is intended to make your collecting more meaningful.

The opening remarks will give you a general idea of mineralogy to which the specific information in respect to the pictured minerals can be related. Any of the descriptive terms used with which you may not be familiar such as drusy, botryoidal, etc., can be found in a good dictionary. If you find your interest growing to the point where you want a more complete understanding of the science of mineralogy you should consult any of the many comprehensive books available.

You probably know that geology is the broad science of the physical and historical aspects of the Earth as revealed to us in rocks. Since the rocks are made up of minerals, the minerals themselves are of prime importance. Basically, then, mineralogy deals with the very elements and compounds of which the Earth's crust is composed. Because of this, other branches of geology depend to a great extent upon mineralogy—for instance, PETROLOGY, the study of rocks, and ECONOMIC GEOLOGY, the study of ore deposits. Without a good knowledge of minerals, a petrologist could hardly determine the origin of rocks, nor could an economic geologist study valuable ore materials.

Approximately 2,000 minerals are known; about 120 are considered common, and the remaining 1,880 are considered rare or very rare! The more common ones make up the bulk of the rocks and occur in large quantities. Many are found only as deposits in small concentrated aggregates, and are therefore valuable. Some are found in small quantities in unusual or selected types of deposits, and are relatively rare.

Generally, minerals are classified according to the chief *ion* (or ion group) which serves as a mineral's fundamental unit. The minerals in this book are grouped in the order of these classes. This list follows that order:

1. NATIVE ELEMENTS: Only one element required to make a mineral. 2. SULFIDES AND SULFOSALTS: Sulfur ion (S) is fundamental ion. 3. HALIDES: Fluorine (F) or chlorine ion (Cl) is fundamental unit. 4. OXIDES AND HYDROXIDES: Oxygen (O) and hydroxyl ion (OH) is fundamental unit. 5. CARBONATES: Carbonate ion (CO_3) is fundamental unit. 6. BORATES: Borate ion (BO_4) is

fundamental unit. 7. SULFATES: Sulfate ion (SO_4) is fundamental unit. 8. CHROMATES, TUNGSTATES, MOLYBDATES: Chromate (CrO_4), Tungstate (WO_4), Molybdate (MoO_4) ions are fundamental units. 9. PHOSPHATES, ARSENATES, VANADATES: Phosphate (PO_4), Arsenate (AsO_4), Vanadate (VO_4) ions are fundamental units. 10. SILICATES: Silicate ion (SiO_4) is fundamental unit.

The most abundant minerals making up *igneous* rocks are the silicates: quartz, feldspars, micas, augite, hornblende, and olivine. In *sedimentary* rocks, clay minerals, quartz, feldspars, calcite, and micas are abundant. However, many other minerals occurring in minor quantities in small deposits often are more interesting, more colorful, and more sought after by collectors.

In this book you will find the most basic descriptive information about each of the pictured minerals, that is, chemical formula, physical properties, crystal system, and most important to you as a collector, a summary of localities where they are most commonly found. Let us take a brief look at the principal physical properties of a mineral—all of which involve very simple hand tests which will be an invaluable aid in your identification of specimens. These are color, streak, lustre, hardness, and specific gravity.

1. COLORS of minerals are pretty fair guides since they are for the most part characteristic. However, color can be deceiving because a very small amount of impurity can have a strong effect in altering the color but not the other properties. Do not overdepend upon color.

2. STREAK is the color of the powder of a mineral.

Test this by rubbing the mineral on a white, hard ceramic plate, or, if necessary, by grinding.

3. Lustre refers to the general surface appearance in reflected light, such as vitreous, adamantine, greasy.

4. Hardness refers to the ease with which a mineral can be scratched and is measured by number on the basis of *Mohs' scale*, which goes from soft to hard as follows: (1) talc; (2) gypsum; (3) calcite; (4) fluorite; (5) apatite; (6) orthoclase; (7) quartz; (8) topaz; (9) corundum; (10) diamond. This means that a mineral that scratches as easily as talc has a hardness of 1, while a mineral with a scratch level of quartz has a hardness of 7. This scale does not represent the *true* hardness relationship of the minerals to each other. For example, diamonds are actually 150 times as hard as corundum, the next hardest mineral on the scale.

5. Specific Gravity. Usually abbreviated Sp. G., this simply means the weight of a mineral as compared to water. For example, a mineral with a Sp. G. of 3 is 3 times heavier than an equal *volume* of water.

Even with these tests, you may be wondering what *really* distinguishes one mineral from another. We now come to the most complex aspect of minerals and one which we cannot possibly deal with at length here— crystal structure. Crystallography is a whole subdivision of mineralogy dealing with the external form of naturally occurring crystals of the different minerals. Identification of minerals is made much easier if well-shaped crystals are available, and many minerals actually do develop such characteristic crystals that identification can be made on sight. The photographs of minerals contained in this book in many cases have been magnified in order to bring out the full color inherent in the mineral as well as to reveal characteristic crystal shapes and forms.

The atoms making up minerals show a very orderly inner geometrical arrangement which gives each mineral *crystallinity*—a property which allows the material to develop into crystals. What is a crystal? Merely a solid body, often displaying plane natural surfaces called *faces*, which reflect the regular internal atomic arrangements. Crystals may be very small, or may be very large, depending upon how often the internal pattern can repeat itself in all three directions, which in turn depends upon the supply of material available to begin with.

Actually, small-sized crystals have a better chance to develop good faces and outlines than large crystals, because complete freedom of growth requires special environmental balance. Growing space is usually limited and crystal growth frequently is hindered. Consequently, minerals commonly grow with poor face development, and well-shaped crystals are relatively uncommon.

You probably have always thought of crystals as having shiny flat faces, but you were wrong. Even if flat external faces are absent, the orderly geometrical arrangement of atoms still exists internally. So do not be deceived by poorly developed external appearances. If you X-ray an amorphous-looking mineral, you will find that it falls into one of six crystal systems:

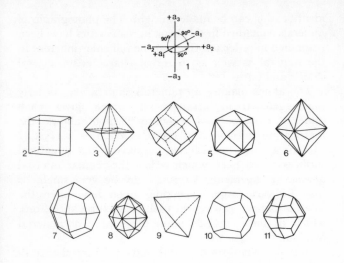

1. *Isometric System:* All crystals in this system have three axes (main directions) of equal length, intersecting at right angles to each other. The axes are denoted a_1 (front to back), a_2 (right to left), and a_3 (vertical). Similar faces on a crystal constitute a *form*. In this system, the most common forms are: cube (2), octahedron (3), dodecahedron (4), tetrahexahedron (5), trisoctahedron (6), trapezohedron (7), hexoctahedron (8), tetrahedron (9), pyritohedron (10), diploid (11).

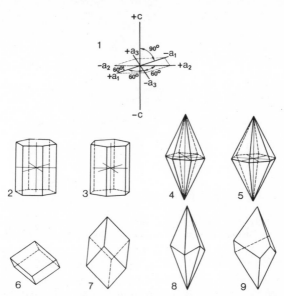

2. *Hexagonal System:* All crystals in this system can be referred to four axes, three of which are horizontal intersecting at 60°; the fourth axis is vertical and at right angles to the other three. The horizontal axes are of equal length, and are designated a_1, a_2, a_3. The vertical axis is not equal to the a axes, and is designated c. Common forms in this system are: base, prism of the first order (2), prism of the second order (3), first order pyramid (4), second order pyramid (5), positive rhombohedron (6), negative rhombohedron (7), scalenohedron (8), trapezohedron (9).

3. *Tetragonal System*: In this system, crystals have two horizontal axes of equal length, a_1 (front to back) and a_2 (right to left) intersecting at right angles, and a third axis (vertical), not equal to a_1 and a_2, designated c. Common forms in this system are: first order pyramid (2), second order pyramid (3), first order prism (4), second order prism (5).

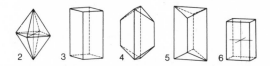

4. *Orthorhombic System:* All crystals with three unequal axes, intersecting at right angles, are placed in this system. The longest axis is usually designated *c* (vertical), while the shortest of the other two is designated *a* (front to back). The other axis, from right to left, is designated *b*. Common forms are: pyramid (2), prism (3), prism and dome (4), prism and dome (5), pinacoid (6).

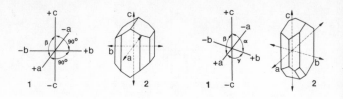

5. *Monoclinic System:* Crystals which are referred to three unequal axes, one of which is inclined, are classed as monoclinic. The axis designated *a* (front to back) is the inclined axis, *b* (horizontal from right to left), and *c* (vertical). The angle between *a* and *c* is necessarily greater than 90°. The angle between *a* and *b* and between *b* and *c* are right angles. Common forms are: dome, prism, and pyramid, usually in combination (2).

6. *Triclinic System:* Crystals which are referred to three unequal axes, all with oblique intersections (greater than 90°), are called triclinic. The *a* axis is from front to back, *b* from right to left, and *c* is vertical. Common forms are: dome, prism, pyramid, usually in combination (2).

Many of the colorful minerals in this book are shown on rocks and with other associated minerals as they occur in nature; others have been separated from the rocks and placed suitably for best photographic effects to aid you in ready identification once you embark on this fascinating journey into the colorful world of minerals.

CONTENTS

NAME: **Copper**   FORMULA: Cu

PHYSICAL PROPERTIES:

Color: Pale red when fresh, tarnishing to brownish shades

Streak: Shining pale red   *Lustre:* Metallic

Hardness: 2.5–3.0   *Sp. G.:* 8.95

CRYSTAL SYSTEM: Isometric. Usually in nodular, sheet-like, or branching masses, on which cubic crystals, displaying cubic dodecahedral and octahedral faces, are rudely assembled.

OCCURRENCES: Often found in the oxidized portions of copper-sulfide ore bodies. In U.S. nicely shaped crystals at New Cornelia, Arizona and Santa Rita, New Mexico. Also found as cavity fillings in basalts and in conglomerates in Keeweenaw Peninsula, Michigan. Also found at Somerville, New Jersey. In South Australia at Wallaroo and on Yorke Peninsula, and in New South Wales at Broken Hill. In U.S.S.R. very fine crystals at Turnisk in Perm and near Nizhne-Tagilsk. Good dendritic and arborescent crystals are found in Cornwall, England. Native copper wires occur at Tsumeb, South West Africa.

Copper

NAME: **Silver**   FORMULA: Ag

PHYSICAL PROPERTIES:

Color: Brilliant white when fresh, but tarnishes easily to black or grey

Streak: Shining silver white   *Lustre:* Metallic

Hardness: 2.5–3.0   *Sp. G.:* 10.5

CRYSTAL SYSTEM: Isometric, but not usually distinctly so. Often in irregular masses or wires; also in sheets or plates. Cubes seen occasionally, as well as octahedrons and dodecahedrons. Plate-like crystals often take on an arborescent or dendritic arrangement.

OCCURRENCES: Found in oxidized portions of hydro-thermal sulfide veins, or as primary minerals in sulfide deposits. In Canada, as sheets and irregular masses in veins at Cobalt, Gowganda, and O'Brien in Ontario; as nice dendritic masses with pitchblende at Great Bear Lake. Also with native copper as cavity-fillings in basalt at Keeweenaw Peninsula, Michigan. Found in Colorado at Aspen and in Boulder County; also in several mining areas in Arizona. Good specimens have been found at the Kongsberg mines in Norway. In Australia, silver is found at Broken Hill in New South Wales. Mexico has been a big producer, particularly in Batopilas, in the state of Chihuahua; also at Guanajuato, and Sonora, Durango, Zacatecas.

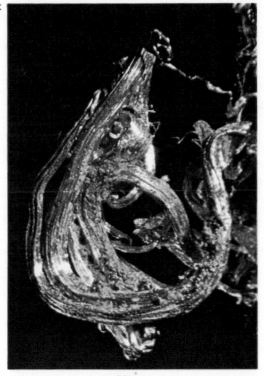

Silver

NAME: **Gold**   FORMULA: Au

PHYSICAL PROPERTIES:

Color: Gold yellow, rarely with orange tint

Streak: Gold yellow, slight white or   *Lustre:* Metallic
orange tint

Hardness: 2.5–3.0   *Sp. G.:* 19.3

CRYSTAL SYSTEM: Isometric. Not often in distinct crystals. Usually arranged in dendritic growths, but masses show rounded cavernous appearance. When in good crystals, gold displays octahedrons, dodecahedrons, and cubes.

OCCURRENCES: Gold is often found in hydrothermal veins as strings, scales, plates, and irregular masses. In the western U.S., splendid masses are found in the Mother Lode country of California, as in El Dorado, Calaveras, Kern and Tuolumne Counties. Fine crystals and masses from Leadville, Silverton, Telluride, Cripple Creek, and Breckenridge districts of Colorado. Also at Homestake Mine, South Dakota; in Montana and Arizona. Also in Ural Mts., U.S.S.R.; and at Waihi Mine, New Zealand. In Australia, at Mount Morgan Mine, from Chartres Towers district, from Hillgrove, from Hill End, and large production from Kalgoorlie and Ballarat. In Britain, near Dolgellan in Wales.

Sulfur

NAME: **Sphalerite**   FORMULA: ZnS

PHYSICAL PROPERTIES:

Color: Usually yellow, brown, black. Often red, green to nearly white. When absolutely pure, nearly colorless

Streak: Brownish to light yellow to white

Lustre: Resinous to adamantine

Hardness: 3.5–4.0

Sp. G.: 3.9–4.1

CRYSTAL SYSTEM: Isometric. Usually tetrahedral, but somewhat resembling octahedrons. Commonly in fine-grained to coarse-grained masses. Frequently found with galena.

OCCURRENCES: In replacement bodies in sediment, but more commonly in hydrothermal deposits. Occasionally in igneous and contact metasomatic deposits. In Germany, found as nicely shaped crystals at Neudorf, Harz Mts.; at Pribram in Czechoslovakia. Also at St. Agnes and elsewhere in Cornwall, England; from Picos de Europa, Spain. Very beautiful crystals, associated with galena, marcasite, and calcite are found at Joplin, Missouri. Fine crystals of dark color at Breckinridge, Colorado. Excellent crystals at Butte, Montana. Green varieties have been found at Franklin, New Jersey. Well-shaped crystals at Binnatal, Switzerland.

Sphalerite

(Grey crystals beneath large Sphalerite crystal is Galena.)

NAME: **Chalcopyrite** FORMULA: $CuFeS_2$

PHYSICAL PROPERTIES:

Color: Brassy yellow, but easily tarnishes to bronze or iridescent bronze, purples, and blues.

Streak: Greenish black   *Lustre:* Metallic, but often tarnished

Hardness: 3.4–4.0   *Sp. G.:* 4.1–4.3

CRYSTAL SYSTEM: Tetragonal, often resembling tetrahedrons. When small, crystals are usually very well formed, but become irregular with increase in size. Often drusy, but also as massive growths of crystals.

OCCURRENCES: Usually found in hydrothermal deposits in veins; also in contact metasomatic deposits. Found in nicely shaped crystals scattered on dolomite in Joplin area of Missouri. Fairly large crystals in Clear Creek County, Colorado. Very fine crystals in dolomite from French Creek Mines, Pennsylvania. Also at La Bufa, Mexico. Very unusual crystals at Carn Brea Mine, Cornwall, England. Found at Wanlockhead Mine in Scotland, and also at Wallaroo, South Australia.

5 ×

Chalcopyrite

NAME: **Galena**   FORMULA: PbS
PHYSICAL PROPERTIES:
Color: Blue-grey when fresh, to pale dark grey
Streak: Lead grey   *Lustre:* Metallic
Hardness: 2.5   *Sp. G.:* 7.6
CRYSTAL SYSTEM: Isometric. Very common in cubes or cubes with octahedrons. Sometimes twinned. Usually in massive aggregates of cubic crystals. Faces often brightly shiny and smooth, occasionally dull. Occasionally fibrous-like masses.
OCCURRENCES: Very common in hydrothermal deposits, and often associated with pyrite, sphalerite, and chalcopyrite. In U.S., common in Oklahoma-Kansas-Missouri area where hydrothermal solutions deposited galena in cavities in chert-limestone. Excellent crystals in this area range up to 6″ in size, but commonly 1″–2″. Splendid cubes from Leadville, Colorado. Also at Couer d'Alene, Idaho, and Tintic, Utah. Beautiful crystals in Saxony, Germany. Massive crystals at Broken Hill, New South Wales, Australia. Twinned specimens are found in Cornwall, England.

Galena
(Yellow crystals beneath are probably Barite.)

NAME: **Cinnabar**   FORMULA: HgS

PHYSICAL PROPERTIES:

Color: Scarlet red to slightly brownish red

Streak: Scarlet red   *Lustre:* Adamantine

Hardness: 2.0–2.5   *Sp. G.:* 8.1

CRYSTAL SYSTEM: Hexagonal, but crystals are usually rhombohedral; also thick and tabular. Often in drusy crystals coating others. Also acicular prismatic.

OCCURRENCES: Cinnabar is found in low-temperature hydrothermal deposits, associated with volcanic regions. In U.S., fine crystals in calcite at Cahill Mine, Nevada. Also at New Almaden and New Idria Mines, California, as nicely shaped small crystals coating fractures. Bright red crystals also at Charcas, San Luis Potosí, Mexico. Excellent crystals at Almadén, Spain. Found also at Mt. Avala, Belgrade, Yugoslavia. Some of the world's finest cinnabar crystals come from Kweichow and Hunan Provinces, China.

Cinnabar

NAME: **Pyrrhotite**   FORMULA: FeS

PHYSICAL PROPERTIES:

Color: Between bronze yellow and copper red

Streak: Dark greyish black   *Lustre:* Metallic

Hardness: 3.5–4.5   *Sp. G.:* 4.6

CRYSTAL SYSTEM: Hexagonal. Usually thin tabular plates with hexagonal outline. ' Often coated with other crystals. Often found massive and granular.

OCCURRENCES: Found in basic igneous rocks; also in contact metasomatic deposits and veins. Occasionally in pegmatites. Found as irregular masses in basic rocks at Sudbury, Canada; also in 1″ crystals at Bluebell Mine, British Columbia. In pegmatites at Standish, Maine. Large (3″) well-shaped crystals at Trepca, Yugoslavia. Also very large crystals at Llallagua, Bolivia; and from Morro Velho Mines, Minas Gerais, Brazil.

Pyrrhotite

NAME: **Millerite**   FORMULA: NiS

PHYSICAL PROPERTIES:

Color: Brassy to bronzy yellow

Streak: Greenish black   *Lustre:* Metallic

Hardness: 3.0–3.5   *Sp. G.:* 5.5

CRYSTAL SYSTEM: Hexagonal; occurs usually as long, slender, prismatic filaments, terminated with a low rhomb. Found as radiate sprays and delicate matted fibrous masses. Occasionally granular.

OCCURRENCES: Found in low-temperature hydrothermal deposits associated with other nickel minerals. In U.S., good crystals found in geodes near Keokuk, Iowa, and in cavities in limestones in Mississippi Valley area. Nice crystals at Sieger, Westphal, Germany; at Kladno-Rodna, Czechoslovakia. Very unusual but nice cleavable large crystals occur at Timagami, Ontario, Canada. It has been found as fine hair-like crystals at Merthyr Tydfil, Wales.

Millerite

NAME: **Stibnite**   FORMULA: Sb_2S_3

PHYSICAL PROPERTIES:

Color: Pale to dark lead grey, to steel grey

Streak: Grey to dark grey   *Lustre:* Metallic

Hardness: 2.0   *Sp. G.:* 4.6

CRYSTAL SYSTEM: Orthorhombic. Usually long prismatic crystals, with striated or furrowed faces, terminated by pyramidal faces. Also as aggregates of radiating groups.

OCCURRENCES: Found in low-temperature hydrothermal veins with other sulfides. In U.S., nice bladed crystals occur in the Darwin district, Inyo County, California; also at Antimony Peak, San Benito County, California; and large crystals come from the Rand district, San Bernardino County. Long crystals at Manhattan Mines, Nye County, Nevada. Excellent needle-like crystals come from Felsobanya, Rumania. A good source of stibnite mixed with realgar and cinnabar is Pereta, Italy. Several localities in France produce stibnite, also. Excellent crystals come from Macedonia, Greece. The most famous locality is the Ichinokowa area, Shikoku Island, Japan.

Stibnite

NAME: **Pyrite**   FORMULA: FeS_2

PHYSICAL PROPERTIES:

Color: Pale brassy yellow

Streak: Greenish to brownish black   *Lustre:* Metallic

Hardness: 6.0–6.5   *Sp. G.:* 5.0

CRYSTAL SYSTEM: Isometric. Usually found as cubes and pyritohedrons, with octahedrons sometimes. Crystals often in clusters, druses, and also isolated individuals. Parallel striations on cubes and pyritohedrons.

OCCURRENCES: A very widespread mineral, the most common of the sulfides. Found abundantly in hydrothermal deposits and in contact metasomatic deposits. In U.S., very large crystals, up to 5″ across, come from Bingham, Utah. Brilliant well-shaped crystals, up to 1″ across, are present at Park City, Nevada. In Colorado, good specimens come from Rico, Gilman, and Leadville. Very fine octahedral forms occur at French Creek Mines, Pennsylvania. Pyrite is common also in Mexico, Bolivia, and Peru. Large fine crystals are recorded at Rio Marina, Italy. Fine crystals in Cornwall and Devonshire, England.

Pyrite

NAME: **Marcasite**   FORMULA: FeS_2
PHYSICAL PROPERTIES:
Color: Pale brassy yellow to greyish green
Streak: Greyish to brownish black   *Lustre:* Metallic
Hardness: 6.0–6.5   *Sp. G.:* 4.9
CRYSTAL SYSTEM: Orthorhombic. Usually in tabular, prismatic crystals in aggregates. Crystals often show pyramidal faces. Seen in radiating fibres. Also in massive and cockscomb forms.
OCCURRENCES: Found in low-temperature environments, in sediments as concretions and encrustations. Also in sulfide veins associated with lead and zinc minerals. In U.S., very abundant in Tri-State lead-zinc district as cockscomb forms. Very excellent cockscomb crystals are found in Racine County, Wisconsin; at Red Bank, New Jersey; at Galena, Illinois. Also found in Czechoslovakia; in France; at Dover Cliffs and in Devonshire, England; in Mexico.

3 ×

Marcasite

39

NAME: **Arsenopyrite**   FORMULA: FeAsS

PHYSICAL PROPERTIES

Color: Silver-white to greyish white

Streak: Dark greyish black   *Lustre:* Metallic

Hardness: 5.5–6.0   *Sp. G.:* 5.5–6.0

CRYSTAL SYSTEM: Orthorhombic. Usually in well-developed crystals showing prisms. Sometimes striated. Seen often as granular masses.

OCCURRENCES: Found in high-temperature mineral deposits, associated with sulfides, and scattered in metamorphic rocks. In the U.S., nicely shaped crystals occur at Franconia, New Hampshire, and at Franklin, New Jersey. Also, at Leadville, Colorado; and associated with gold at Homestake, South Dakota. Good crystals 2″ across at Hidalgo de Perral, Mexico. Also, 1″–2″ crystals are found at Llallagua, Bolivia. Also, in Norway and in Cornwall and Devonshire, England. Fine crystals are found in schists at Mitterberg, Austria, and in the Trentino district, Italy.

Arsenopyrite
(Metallic silvery crystals; brassy-yellow
crystals probably Chalcopyrite; white
mass in foreground probably Barite.)

NAME: **Realgar**   FORMULA: AsS

PHYSICAL PROPERTIES:

Color: Aurora red to orange-yellow

Streak: Orange-red to aurora red *Lustre:* Resinous

Hardness: 1.5–2.0   *Sp. G.:* 3.5

CRYSTAL SYSTEM: Monoclinic. Prismatic crystals usually with vertical striations; but often short and stubby. Sometimes found as granular masses, and very compact masses.

OCCURRENCES: Realgar is usually found in low-temperature hydrothermal deposits, associated with other arsenic minerals, particularly yellow-orange orpiment, and with other sulfides. In U.S., good specimens of large size occur at Getchell Mine, Nevada. Pretty crystals occur at Manhattan, Nevada, and also at Boron, California. Some fine specimens occur at Mercur, Utah. Some are found around hot springs in Yellowstone Park. Also at Binnatal, Switzerland; at Nagyag, Rumania.

Realgar
(Long red to orange crystals; small lemon-
yellow crystals are Orpiment.)

NAME: **Orpiment**   FORMULA: As_2S_3

PHYSICAL PROPERTIES:

Color: Lemon yellow to yellow-orange

Streak: Pale lemon yellow   *Lustre:* Resinous to pearly

Hardness: 1.5–2.0   *Sp. G.:* 3.5

CRYSTAL SYSTEM: Monoclinic. Crystals usually poorly developed into thick columnar masses and wedges.

OCCURRENCES: Orpiment is found usually in low-temperature hydrothermal deposits, often associated with orange-red realgar. In U.S., well-shaped large crystals found at Mercur, Utah, and at the Getchell Mine, Nevada. Quite beautiful masses occur at Manhattan, Nevada. Good aggregates at Morococha, Peru; also at Tajowa, Hungary; also in Yugoslavia. Very fine specimens come from Allchar, Greece, and from Balin, Turkey.

5 ×

Orpiment

(Lemon-yellow mass; with red-orange Realgar.)

NAME: **Tetrahedrite-**   FORMULA: Cu_3SbS_3
Tennantite Cu_3AsS_3

PHYSICAL PROPERTIES:

Color: Flint grey to iron black

Streak: Flint grey to black,   *Lustre:* Metallic
sometimes brownish grey

Hardness: 3.0–4.0   *Sp. G.:* 4.6–5.1

CRYSTAL SYSTEM: Isometric. Usually in tetrahedrons; also in modified cubes and dodecahedrons. Crystals often form druses on sedimentary rock. Sometimes as granular masses.

OCCURRENCES: In copper or silver hydrothermal veins and in contact metamorphic deposits. Beautiful sharp crystals have been found at Bingham Canyon, Utah. Very fine small crystals occur at Kellogg Mine, Idaho. Another U.S. locality is Butte, Montana. Very good crystals have been found at Botés, Rumania, and at Pribram, Czechoslovakia. In Germany, very excellent crystals at several places in Harz Mts., particularly Neudorf. From copper-tin veins in Cornwall, England, and from Potosí, Bolivia.

Tetrahedrite-Tennantite
(Coated with tiny Chalcopyrite crystals.)

NAME: **Proustite**   FORMULA: Ag_3AsS_3

PHYSICAL PROPERTIES:

Color: Bright red to scarlet

Streak: Bright red   *Lustre:* Adamantine to submetallic
  to scarlet

Hardness: 2.0–2.5   *Sp. G.:* 5.55

CRYSTAL SYSTEM: Hexagonal. Usually rhombohedrons
are displayed. Found also massive to compact. Also,
good prismatic forms develop, and often are striated.

OCCURRENCES: Found in hydrothermal silver veins,
associated with other silver minerals. In U.S., proustite
occurs at silver mining areas of the Rocky Mts., such as
at Ruby district, Colorado; in Comstock Lode area of
Nevada; at Silver City, Idaho; at Silverton, Colorado.
Very fine crystals occur at Dolores Mine, Chanarcillo,
Chile. Splendid crystals are present at Joachimsthal,
Czechoslovakia, and in Germany at various silver
districts in Saxony, such as at Annaberg and Freiberg.

Proustite

NAME: **Pyrargyrite**   FORMULA: Ag_3SbS_3

PHYSICAL PROPERTIES:

Color: Black to greyish black with dark red tone

Streak: Dark purplish red   *Lustre:* Metallic to adamantine

Hardness: 2.5   *Sp. G.:* 5.85

CRYSTAL SYSTEM: Hexagonal. Hexagonal prisms with blunt pyramidal ends are common, but usually not well developed. Seen in interlocked form as aggregates of crystals. Also compact.

OCCURRENCES: Pyrargyrite usually forms in low-temperature hydrothermal deposits, often with proustite, argentite, and other silver-bearing minerals. Pyrargyrite occurs at silver districts in the western U.S., as at Ruby district in Colorado; Silver City, Idaho; at Silverton, Colorado. Very fine specimens are found at Colquechaca, Bolivia. Pyrargyrite, associated with proustite, occurs in Atacama, Chile. Very fine crystals at Andreasburg, in Harz Mts., Germany. Quite nice prismatic crystals were formerly found in silver mines at Guanajuato, Mexico.

Pyrargyrite

NAME: **Bournonite**   FORMULA: $PbCuSbS_3$

PHYSICAL PROPERTIES:

Color: Steel grey to blackish lead grey

Streak: Steel grey to black   *Lustre:* Metallic

Hardness: 2.5–3.0   *Sp. G.:* 5.8

CRYSTAL SYSTEM: Orthorhombic. Stubby prismatic crystals, often slightly tabular, and often terminated by other prisms. Commonly striated by twinning which develops crude cross form.

OCCURRENCES: Bournonite is usually found in hydrothermal deposits associated with galena, sphalerite, stibnite, and other sulfides. Good crystals are not too common. In U.S., fair crystals of large size are found at Park City, Utah; also found in Arizona and at Emery, Montana. Small crystals occur at Zacatecas, Mexico. Brilliant large crystals come from the Vibora Mine, Bolivia. Very nice specimens are found at Endellion and Liskeard, Cornwall, England. Nice 2″ crystals occur at Andreasberg, Harz Mt. area, Germany. Fine, sharp, well-shaped crystals come from the Nakaze Mine, Japan.

Bournonite

NAME: **Halite**   FORMULA: NaCl

PHYSICAL PROPERTIES:

Color: Colorless or white, often tinted blue, pink, or yellow

Streak: Colorless   *Lustre:* Vitreous

Hardness: 2.5   *Sp. G.:* 2.17

CRYSTAL SYSTEM: Isometric. Cubes very common, sometimes with octahedrons. Faces smooth. Also massive.

OCCURRENCES: Very widespread in saline sediments. In U.S., found at Searles Lake, California. Good crystals in many counties of New York State. Salt beds in Ohio and Michigan yield fine crystals. Nice crystals in Louisiana in salt mines. Also present in Kansas. Salt mines at Iletskaya Zashchita in Siberia, U.S.S.R., yield excellent crystals. Very extensive artificial caverns at Wieliczka, Poland, contain beautiful and large cubic crystals. Halite of blue color comes from Stassfurt, Germany. Large deposits occur in Cheshire, England.

Halite
(Both forms.)

Fluorite

× 1/3

NAME: **Fluorite**   FORMULA: CaF_2

PHYSICAL PROPERTIES:

Color: Colorless, white, yellow, green, red, violet-blue, brown

Streak: White

Hardness: 4.0

Lustre: Vitreous

Sp. G.: 3.18

56

Fluorite
(Pink
crystals;
with
Quartz
crystals.)

3 ×

CRYSTAL SYSTEM: Isometric. Cubes very common, often modified with octahedrons. Sometimes massive forms and granular aggregates occur.

OCCURRENCES: Fluorite occurs in a variety of types of deposits: hydrothermal, sedimentary, igneous, and volcanic. Very fine specimens are found with amethyst in lead-silver mines in the Thunder Bay district, Canada. Sharp, well-shaped crystals occur at Madoc, Marmora, and Huntingdon, Ontario, Canada. In U.S., excellent crystals, green in color, are found at Westmoreland, New Hampshire. At Clay Center, Ohio, limestones contain nice cubic, yellow-brown crystals. The best known locality in the U.S. is at Rosiclare, Illinois, where large violet cubes occur. Fine green specimens are found at Weardale, Durham, England. Other English localities include Alston Moor, Cleator Moor, in Cumberland, and the Wheal Mary Mine in Cornwall. Yellow crystals come from Gersdorf, Germany.

NAME: **Atacamite**   FORMULA: $Cu_2(OH)_3Cl$

PHYSICAL PROPERTIES:
Color: Bright green to dark green
Streak: Apple green   *Lustre:* Adamantine to vitreous
Hardness: 3.0–3.5   *Sp. G.:* 3.8

CRYSTAL SYSTEM: Orthorhombic. Usually as long, slender prismatic crystals, sometimes vertically striated. Crystals aggregated in clusters. Sometimes massive.

OCCURRENCES: Atacamite is always formed by secondary processes from copper-bearing minerals by weathering. In U.S., very nice crystals are found at San Manuel, Arizona; but rare in Wyoming, and rare in Utah. Fine specimens come from Chuquicamata, Chile. At Ravensthorpe, Western Australia, and at Wallaroo, South Australia, very good pale green specimens have been found.

10 X

Atacamite
(Long, green crystals; yellow mass probably
fine-grained Gypsum.)

NAME: **Cuprite**   FORMULA: Cu_2O

PHYSICAL PROPERTIES:

Color: Red, varying to dark red

Streak: Brownish red   *Lustre:* Adamantine to submetallic

Hardness: 3.5–4.0   *Sp. G.:* 6.1

CRYSTAL SYSTEM: Isometric. Often as octahedrons, also as cubes and dodecahedrons. Occasionally in long fibrous crystals; also massive or granular.

OCCURRENCES: Cuprite is usually of secondary origin, forming by oxidation of copper minerals. Common in copper terranes. In U.S., good specimens come from Bisbee, Morenci, and Globe, Arizona. Nice crystals about $\frac{1}{4}''$ across are found at Santa Rita, New Mexico. Fairly large $1''$ crystals at Bingham, Utah. Also, formerly in the Perm area at Bogoslovsh, U.S.S.R. Very splendid single crystals come from Chessy near Lyon, France. Good crystals in sulfide veins near St. Day, Cornwall, England. Nice crystals occur at Burra District, South Australia, at Broken Hill, New South Wales, and at Mt. Isa, Queensland.

Cuprite
(Red crystals; with white carbonate crystals.)

NAME: **Magnetite**   FORMULA: Fe_3O_4

PHYSICAL PROPERTIES:

Color: Iron black

Streak: Black   *Lustre:* Submetallic to metallic, sometimes dull

Hardness: 5.5–6.5   *Sp. G.:* 5.2

CRYSTAL SYSTEM: Isometric. Usually as octahedrons; occasionally dodecahedrons, too. Cubes rare. Also massive; granular.

OCCURRENCES: Magnetite is a fairly abundant mineral, being found in crystalline rocks, in igneous and metamorphic terrane; also in contact-metamorphic areas. In U.S., very fine 1″ specimens occur at Mineville, Essex County, and in Tilly Foster Iron Mine in Putnam County, New York. Very beautiful crystals occur at French Creek Mines, Pennsylvania. Large magnetite crystals are found in Iron County, Utah, and at Cornwall, Pennsylvania. Large crystals occur also at Faraday, Ontario, Canada. Well-shaped crystals are found at Nordmark, Sweden; also at Zillertal and Pfischtal, Austria.

Magnetite
(Black octahedrons in fine-grained bedrock.)

OXIDES AND HYDROXIDES

NAME: **Hematite**   FORMULA: Fe_2O_3

PHYSICAL PROPERTIES:

Color: Steel grey to black; red when earthy

Streak: Deep red to   *Lustre:* Metallic to submetallic;
reddish brown   also earthy

Hardness: 5.5–6.0   *Sp. G.:* 5.3

CRYSTAL SYSTEM: Hexagonal. Rhombohedral crystals common; often flat tabular crystals in "iron roses." Also columnar, and quite often in rounded bubbly-like masses.

OCCURRENCES: Found in a great many types of deposits, including volcanic rocks, deep-seated igneous rocks, in

Hematite

metamorphic rocks, and also in rocks formed by weathering processes. In U.S., in nicely shaped crystals at Edwards and Antwerp, New York. Very fine 4″ rhombohedrons occur at Franklin, New Jersey. Clusters of fine crystals are present at Twin Peaks, Utah. Found in rounded botryoidal form at iron mines in Michigan and Minnesota. Good octahedrons (martite) at Durango, Mexico. Hematite roses occur in Switzerland. Also in U.S.S.R. and Romania. Nice clusters at Cleator Moor and elsewhere in Cumberland, England, and at Bimbowrie, South Australia.

NAME: **Rutile**   FORMULA: TiO_2

PHYSICAL PROPERTIES:

Color: Reddish brown to dark brown-red

Streak: Pale brown   *Lustre:* Adamantine to metallic

Hardness: 6.0–6.5   *Sp. G.:* 4.2

CRYSTAL SYSTEM: Tetragonal. Usually stubby prismatic; often slender acicular. Often twinned.

OCCURRENCES: Rutile is fairly widespread, occurring in many kinds of rocks and mineral deposits, ranging from igneous to metamorphic to sedimentary. In U.S., beautiful crystals occur on Graves Mt., Georgia. Nice black crystals up to 1″ long are found at Magnet Cove, Arkansas. Very fine specimens occur in pegmatites in North Carolina. Splendid crystals in Inyo County, California. Fine 1″ prismatic crystals at Lofthus, Norway; brilliant crystals at many areas in Alps, particularly Tavetschtal and Binnatal. Splendid crystals at Cerrado Frio, Minas Gerais, Brazil.

Rutile
(Long reddish-brown crystals in Quartz.)

NAME: **Anatase** FORMULA: TiO_2

PHYSICAL PROPERTIES:

Color: Browns passing to indigo blue; also black

Streak: Uncolored   *Lustre:* Adamantine to metallic

Hardness: 5.5– 6.0   *Sp. G.:* 3.9

CRYSTAL SYSTEM: Tetragonal. Usually octahedral in habit; also tabular; rarely prismatic.

OCCURRENCES: Anatase is a fairly rare mineral, and is usually of secondary origin, derived from alteration of titanium minerals. In U.S., fine blue crystals have been found in igneous rocks on Beaver Creek, Gunnison County, Colorado. Also at Somerville, Massachusetts. A few very fine specimens occur at Magnet Cove, Arkansas. Well-shaped crystals at Binnatal and St. Gotthard, Switzerland; at Bourg d'Oisons, France. Gem quality crystals at Minas Gerais, Brazil.

3 ×

Anatase
(Large black crystals; with bluish-white
(?) Apatite crystals.)

NAME: **Brookite**   FORMULA: TiO_2

PHYSICAL PROPERTIES:

Color: Brown; also yellowish to reddish brown

Streak: Uncolored or slightly   *Lustre:* Metallic to
greyish to yellowish   adamantine

Hardness: 5.5–6.0   *Sp. G.:* 3.8–4.0

CRYSTAL SYSTEM: Orthorhombic; often in pyramidal forms and sometimes tabular crystals.

OCCURRENCES: Brookite is rare, like anatase. Usually of secondary origin by alteration of titanium minerals. In U.S., nice black to brown crystals found at Magnet Cove, Arkansas. A few have been found at Somerville, Massachusetts. Formerly found near Ural Mts. in U.S.S.R. Good tabular crystals at Maderanertal, Switzerland and also at Pragatten, Austria. Fairly nice specimens at Tremadoc, Wales.

Brookite
(Brown crystals; in rock material.)

NAME: **Cassiterite**   FORMULA: SnO_2

PHYSICAL PROPERTIES:

Color: Brown to black

Streak: White to greyish   *Lustre:* Adamantine
white to brownish

Hardness: 6.0–7.0   *Sp. G.:* 7.0

CRYSTAL SYSTEM: Tetragonal. Often in prismatic crystals with low pyramidal terminations. Sometimes very slender. Sometimes reniform, massive, or granular.

OCCURRENCES: Cassiterite is usually a high-temperature hydrothermal mineral in granite areas, and in pegmatites. Also fairly abundant in placer deposits. Good crystals are not common in North America, but in the U.S., some occur in Lander County, Nevada, and in Sierra County, New Mexico. Fine, sharp, small crystals in pegmatites in New England, in Virginia, and in San Diego County, California. A few well-shaped crystals in tin deposits at Silver Hill, Virginia. Excellent crystals at Llallagua Araca, and Oruro areas of Bolivia. Large black crystals occur at Zinnwald, Czechoslovakia. Choice small-sized aggregates occur at Emmaville, New South Wales, and at Rex Hill, Tasmania. Deposits in Cornwall, England, now almost exhausted.

Cassiterite
(Large twinned crystals; on rock material.)

NAME: **Goethite**   FORMULA: $HFeO_2$

PHYSICAL PROPERTIES:

Color: Yellow, reddish, blackish brown

Streak: Brownish yellow to ochre yellow   *Lustre:* Submetallic to adamantine to dull

Hardness: 5.0–5.5   *Sp. G.:* 3.3–4.3

CRYSTAL SYSTEM: Orthorhombic. Often prismatic crystals, but also flattened. Also fibrous, massive, reniform, or stalactitic. Often arranged in clusters.

OCCURRENCES: A very widespread mineral, occurring as an alteration mineral from other iron-bearing minerals. In the U.S., in the Lake Superior iron districts usually in large compact masses, especially near Marquette, Michigan. Large 2″, bladed crystals in pegmatites in Crystal Peak area, Colorado. Very nice pseudomorphs of goethite after pyrite occur at Pelican Point, on west shore of Utah Lake, Utah. Fine stalactitic forms are present at Horhausen and Westerwald, Germany. Very attractive aggregates occur at Pribram, Czechoslovakia. Small crystals at Bottalack and St. Just, Cornwall, England. Also at Minas Gerais, Brazil, as orange-yellow clustered aggregates.

Goethite

NAME: **Limonite**   FORMULA: $2Fe_2O_3 \cdot 3H_2O$

PHYSICAL PROPERTIES:

Color: Brown of various shades; also yellowish brown to
  ochre yellow

Streak: Yellowish brown   *Lustre:* Silky; also submetallic,
  dull-earthy

Hardness: 5.0–5.5   *Sp. G.:* 3.6–4.0

CRYSTAL SYSTEM: Not crystalline—it's an amorphous
equivalent of Goethite, with variable water content.
(Much of the material formerly called Limonite is now
thought to be cryptocrystalline Goethite.) Usually
stalactitic or botryoidal or mammillary forms, with
fibrous substructure. Also massive, earthy, concretionary.

OCCURRENCES: Limonite is always produced from
alterations of other iron-bearing minerals. It is a very
widespread mineral, formed under low temperatures
and pressures in a great variety of places where iron-
bearing solutions may gather. Fairly abundant in U.S.
Found as beds at Salisbury and Kent, Connecticut, and
in Berkshire County, Massachusetts. Also in Pennsylvania at several localities. Mined for iron ore in
southern Missouri and in northeast Texas. Also mined
in Bavaria, in Harz Mts., and in Sweden. Good specimens at Rosenau, Czechoslovakia; at Freiberg, Saxony,
Germany; also at Lanlivery and St. Just, Cornwall,
England.

76

3 X

Limonite
(Stalactitic brownish-green masses; also ochre-yellow masses of either Limonite or Goethite.)

NAME: **Manganite**   FORMULA: $MnO(OH)$
PHYSICAL PROPERTIES:
Color: Dark steel grey to iron black
Streak: Reddish brown   *Lustre:* Submetallic
   to nearly black
Hardness: 4.0   *Sp. G.:* 4.3
CRYSTAL SYSTEM: Monoclinic. Often in short striated
prismatic crystals with wedge-shaped or flat terminations.
Also columnar, arranged in fibrous-like clusters.
OCCURRENCES: Manganite is a low-temperature mineral
of hydrothermal deposits, but also occurs in other types
of deposits. In U.S., small aggregates are found at
Jackson Mine, Michigan. Very nice crystals at Wood-
stock, Virginia, also at Vesuvius Mine, at Midvale
Mine, and at other localities in Virginia. Also some at
Picton County, Nova Scotia, Canada. A very famous
locality for obtaining large crystals with quartz is
at Ilfield, Harz, Germany. Also at St. Just, Cornwall,
England. Only a few other places in Europe.

Manganite

NAME: **Calcite**   FORMULA: $CaCO_3$

PHYSICAL PROPERTIES:

Color: Colorless, white; also pale grey, red, green, blue

Streak: Colorless   *Lustre:* Vitreous to subvitreous

Hardness: 3.0   *Sp. G.:* 2.71

CRYSTAL SYSTEM: Hexagonal. Crystal habits variable, but prismatic scalenohedral (long, tapered) and rhombohedral (blocky) forms are predominant. Rhombohedral forms vary from obtuse to acute, and from thin tabular to long prismatic. Crystals also in disc-like form, compressed along "c" axis. Over 700 different forms of calcite have been observed. Often seen clustered on matrix, with pointed crystals in random directions. Doubly terminated crystals and parallel aggregates are rare. Crystals often twinned in several ways. Cleavage rhombs of calcite are commonly striated due to repeated twinning. Often found as cave onyx (stalactitic).

OCCURRENCES: A very common mineral found in a great variety of rocks. Found in hydrothermal deposits; in caves; in many sedimentary rocks, making up limestones and acting as cement in many sandstones. In U.S., fine "dogtooth" crystals are found in basalts at Paterson, New Jersey; also in Massachusetts, and Connecticut. In fairly brilliantly colored, large (4″) crystals on Keeweenaw Peninsula, Michigan. Present in geodes in Illinois and Missouri and elsewhere. Found also as fine crystals in Oklahoma-Kansas-Missouri lead-zinc region. Also around Bisbee, Arizona. Some optical calcite from

80

1 X

Calcite

Calcite

Calcite

Santa Rosa Mountains, California. Large-sized, optical-quality calcite also in Mexico at Rodeo. Good crystals also at Charcas, Mexico. Well-formed scalenohedral variety at Frizington, Cumberland, England, while at Egremont, Cumberland, nice "butterfly twins" were formerly found. Some crystals now from copper mines at Tsumeb and Grootfontein, South West Africa. The most famous source of beautiful crystals is at Helgustadir, on Eskifjord, Iceland, where large crystals of "Iceland spar" are in basalt.

Calcite
(Both forms. Scalenohedral crystal at right.)

NAME: **Siderite**   FORMULA: $FeCO_3$

PHYSICAL PROPERTIES:

Color: Grey, also brown and brownish red

Streak: White   *Lustre:* Vitreous to pearly

Hardness: 3.5–4.0   *Sp. G.:* 3.8

CRYSTAL SYSTEM: Hexagonal. Crystals usually display good rhombohedral form, whose faces are often curved slightly. Also in granular masses; in globular and compact forms.

OCCURRENCES: Siderite is usually of sedimentary origin, but it also forms in hydrothermal deposits, and even in pegmatites. In U.S., fairly large brown crystals are found at Mine Hill, Connecticut, but these are cleavable masses and not individual crystals. Quite fine crystals occur in the Gilman District, and at Crystal Peak area, Colorado. Stalactitic siderite, coated with small siderite crystals, has been found at Bisbee, Arizona. Very beautiful well-shaped crystals at Morro Velho gold mine, Minas Gerais, Brazil; many at Camborne, Redruth, and St. Austell, in Cornwall, England. Several localities in Germany.

Siderite
(Small reddish-brown crystals; on
(?) Calcite.)

CARBONATES

NAME: **Rhodochrosite**   FORMULA: $MnCO_3$

PHYSICAL PROPERTIES:

Color: Rose red, also grey to brown
Streak: White   *Lustre:* Vitreous to pearly
Hardness: 3.5–4.5   *Sp. G.:* 3.5

10 X

Rhodochrosite

CRYSTAL SYSTEM: Hexagonal. Distinct well-shaped crystals are uncommon, but rhombohedrons are usually displayed. Also occurs as compact to granular masses.
OCCURRENCES: Often of sedimentary origin, but also found in moderate-temperature hydrothermal deposits,

associated with copper, lead, and zinc sulfide minerals. In U.S., very fine, vivid red rhombohedrons sprinkled on matrix are found at the St. Elmo, Murphy, Moose, John Reed, and Eagle mines in Colorado. Various other mines in Colorado contain greenish grey rhodochrosite rhombohedral crystals. Lovely pink crystals occur at Butte, Montana. Nice small crystals at Beierdorf, Germany; also at Kapnik, Romania; also in Catamarca Province, Argentina.

NAME: **Smithsonite**   FORMULA: $ZnCO_3$
PHYSICAL PROPERTIES:
Color: White to grey, often tinted yellow, brown, or green.
Streak: White   *Lustre:* Vitreous to pearly
Hardness: 4.0–4.5   *Sp. G.:* 4.0–4.4
CRYSTAL SYSTEM: Hexagonal. Usually not well crystallized, but rhombohedrons are displayed. Faces sometimes curved. Common occurrence reniform, botryoidal, or stalactitic; also granular incrustations.
OCCURRENCES: Usually a secondary mineral in oxidized parts of hydrothermal sulfide deposits, derived from zinc minerals, chiefly sphalerite. In U.S., good yellow banded crusts on limestone at Yellville, Arkansas. Very nice blue-green smithsonite is found at Kelly, near Magdelena, New Mexico. Also found as stalactitic types in Inglesias district, Italy. Good massive variety at Tsumeb and Grootfontein, South West Africa. Very large sharp yellow crystals at Broken Hill, New South Wales, Australia.

90

Smithsonite

NAME: **Dolomite**   FORMULA: $CaMg(CO_3)_2$

PHYSICAL PROPERTIES:

Color: White, but often slightly tinted; also rose red, green.

Streak: White   *Lustre:* Vitreous to pearly

Hardness: 3.5–4.0   *Sp. G.:* 2.8–3.0

CRYSTAL SYSTEM: Hexagonal. Rhombohedrons very common, with faces slightly curved. Occasionally prismatic; also occasionally flat tabular.

OCCURRENCES: Dolomite occurs in a variety of sedimentary rocks, and also in hydrothermal deposits. Nicely shaped pink crystals are found in cavities in the Niagara limestone in Ontario, Canada, near Lakes Erie, Ontario, and Huron, and over into New York and Ohio, U.S. Very fine crystals have been found in pegmatites near Stony Point, North Carolina. Also common in limestone of lead-zinc deposits in Mississippi Valley. Also found in mines of Colorado and California. Excellent crystals from Morro Velho gold mines, Minas Gerais, Brazil. Also at Traversella, Italy; at Binnatal, Switzerland; at Cumberland, Cornwall, and Isle of Man, England.

Dolomite

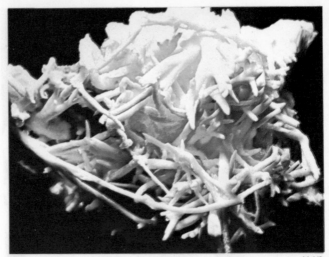

× 1/2

CARBONATES

NAME: **Aragonite** FORMULA: $CaCO_3$

PHYSICAL PROPERTIES:

Color: White; also grey, yellow, or green.

Streak: White *Lustre:* Vitreous to resinous

Hardness: 3.5–4.0 *Sp. G.:* 3.0

CRYSTAL SYSTEM: Orthorhombic. Long prismatic to acicular crystals common, often with acute dome or pyramid terminations. Often globular or coralloidal; also short prismatic. Prismatic crystals appear nearly hexagonal in outline.

Aragonite

1 X

OCCURRENCES: Aragonite is a low-temperature mineral, occurring mainly in deposition with gypsum from saline solutions. In U.S., very fine crystals are found in Magdelena District, New Mexico; also at Las Animas, Colorado. Beautiful delicate aggregates of acicular crystals at Frizington, Cumberland, England. Abundant coralloidal aragonite at Carinthia, Austria; also at Werfen and Erzberg, Austria. Well-developed twins occur at Molina de Aragon, Spain. Splendid transparent yellow crystals up to 3″ in size are found in veins in basalt at Spitzberg, Czechoslovakia. Interesting twins at several places in Sicily.

NAME: **Cerussite**   FORMULA: $PbCO_3$

PHYSICAL PROPERTIES:

Color: White to grey, often tinged blue or green

Streak: White   *Lustre:* Adamantine

Hardness: 3.0–3.5   *Sp. G.:* 6.5

CRYSTAL SYSTEM: Orthorhombic. Stubby to long prismatic crystals common, modified with pyramids. Sometimes tabular; often in floral or star-like aggregates (as shown). Also granular massive.

OCCURRENCES: Cerussite is a low-temperature mineral secondarily produced from galena. Usually found in oxidized portion of lead veins. In U.S., found as large heart-shaped twins at Organ District, New Mexico. Very fine twins in clusters at the Mammoth-St. Anthony mine at Tiger, at the Flux mine, and at Bisbee, Arizona. Found at several places in Germany. Large single crystals and reticulated masses of twins at Tsumeb, South West Africa. Very fine large (6″) crystals at Broken Hill, New South Wales, Australia. Some at Leadhills, Lanark, Scotland.

3 X

Cerussite

NAME: **Azurite**   FORMULA: $Cu_3(OH)_2(CO_3)_2$
PHYSICAL PROPERTIES:
Color: Azure blue
Streak: Pale blue   *Lustre:* Vitreous
Hardness: 3.5–4.0   *Sp. G.:* 3.8
CRYSTAL SYSTEM: Monoclinic. Occurs in a variety of habits. Found as sharp-edged tabular crystals, but also long prismatic or columnar shapes. Prisms usually displayed.
OCCURRENCES: Azurite is a secondary mineral found in oxidized portions of copper-sulfide deposits, both in vein type and disseminated type. In U.S., very nicely shaped crystals are found at Bisbee, Arizona, often in clusters, aggregates, and rosettes. Fairly good crystals of small size are seen at Blue Ball Mine, near Globe, Arizona. Also at San Carlos Mine, Mexico. Very large (2″–3″) and well-shaped azurite crystals are found at Tsumeb, South West Africa. Very nice crystals at several localities in U.S.S.R.; also at Wallaroo, South Australia, and at Broken Hill, New South Wales, Australia.

5 ×

Azurite

99

CARBONATES

20 ×

NAME: **Malachite** FORMULA: $Cu_2(OH)_2CO_3$

PHYSICAL PROPERTIES:

Color: Bright green

Streak: Pale green *Lustre:* Adamantine, toward vitreous

Hardness: 3.5–4.0 *Sp. G.:* 3.6–4.0

CRYSTAL SYSTEM: Monoclinic. Crystals usually in slender, acicular prisms, arranged in clusters, rosettes, or similar groups. Often massive, but also botryoidal rounded structure.

OCCURRENCES: Malachite is a secondary mineral found in the oxidized upper portions of copper-bearing sulfide

Malachite
(Botryoidal; on fine-grained limestone.)

veins and disseminated deposits. It is nearly always associated with azurite. It also serves as an ore mineral of copper. In U.S., good crystals and masses are found at Bisbee and at Morenci, Arizona. Found also at Tintic, Utah, and in New Mexico, and Nevada. A popular European locality is near Sverdlovsk, Russia; at Chessy, France. Excellent massive malachite occurs at Tsumeb, South West Africa; also found at Burra Burra, Australia. In England, near Liskeard, Cornwall, and Alderley Edge, Cheshire.

101

NAME: **Barite**   FORMULA: BaSO$_4$

PHYSICAL PROPERTIES:

Color: White, often tinted yellow, pink, blue, or brown

Streak: White   *Lustre:* Vitreous to resinous

Hardness: 2.5–3.5   *Sp. G.:* 4.3–4.6

CRYSTAL SYSTEM: Orthorhombic. Usually tabular and clustered. Also prismatic, with prisms displayed. Sometimes fibrous; and occasionally granular. Barite "rosettes" are called "desert roses."

OCCURRENCES: Usually a low-temperature hydrothermal vein mineral, but also in sedimentary rocks in veinlets. In Canada, very fine yellow tabular crystals are found at Grand Forks, British Columbia, and in Nova Scotia at the Londonderry Mines. In U.S., nice tabular crystals at several places in St. Lawrence County, New York, particularly Pillar Point. Fine blue tabular crystals are seen at Sterling and at Hartsel, Colorado. Bright yellow specimens occur at Gilman, Colorado. Excellent crystals come from the Dufton mines in Westmorland, England. Other English localities include Cumberland, Shropshire and Derbyshire. Good yellow crystals at Bamle, Norway; in Germany, colorless crystals at Ober-Ostern and from Ilfield, Harz.

× 1/2

Barite

NAME: **Celestite**   FORMULA: $SrSO_4$
PHYSICAL PROPERTIES:
Color: Colorless to white, often tinted blue
Streak: White   *Lustre:* Vitreous
Hardness: 3.0–3.5   *Sp. G.:* 3.9
CRYSTAL SYSTEM: Orthorhombic. Usually tabular or long prismatic, modified with pyramidal forms, arranged in clusters. Also fibrous and radiating.
OCCURRENCES: Celestite is nearly always associated with limestone, but occasionally in gypsum and salt beds, frequently filling veins and other openings. Occasionally found in hydrothermal deposits. In U.S., good blue prismatic $\frac{1}{2}$ to 1″ in size are found in geodes in Lockport dolomite near Syracuse and Rochester, New York, especially near Chittenango Falls. Very fine crystals on cavity walls at Clay Center, Ohio, and nice blue crystals at Portage, Ohio. Large crystals are present at Lampasas, Texas; also at Emery, Utah. Excellent clear crystals near Bristol, England, and in large veins in Gloucestershire; also from Gembock, Germany. Blocky crystals from Mokattam, near Cairo, Egypt.

Celestite

3 ×

SULFATES

NAME: **Gypsum**   FORMULA: $CaSO_4.2H_2O$

PHYSICAL PROPERTIES:

Color: White, occasionally grey, yellow, brown
Streak: White   *Lustre:* Pearly and shining, also vitreous
Hardness: 1.5–2.0   *Sp. G.:* 2.3

CRYSTAL SYSTEM: Monoclinic. Usually in simple mono-clinic crystals, somewhat tabular with good display of prisms and domes. Often prismatic or acicular. Very frequently twinned in swallow-tails.

OCCURRENCES: Gypsum is a very common sulfate, occurring in massive beds in sedimentary areas. It is

106

Gypsum

(Clear crystal; lying on yellow (?) Stalactitic Limonite.)

usually deposited from seawaters and saline lake waters. It is very widespread. Often it occurs as single crystals or twins in clay beds. In U.S., very fine crystals are found in cavities of dolomite in Niagara County, New York. Large crystals up to 2″ in size have been found at Ellsworth, Ohio. Crystals with brown colors in radiating clusters occur at Jet, Oklahoma. Also at Fremont River Canyon, Utah. Good crystals at Cave of Swords, Mexico. Fine crystals at various places in Essex, Oxfordshire and Cheshire, England, and satin-spar variety at Matlock, Derbyshire, England. Also in Germany and Chile.

CHROMATES, TUNGSTATES, MOLYBDATES

NAME: **Crocoite** FORMULA: $PbCrO_4$

PHYSICAL PROPERTIES:
Color: Bright hyacinth red
Streak: Orange-yellow   *Lustre:* Adamantine to vitreous
Hardness: 2.5–3.0   *Sp. G.:* 5.9–6.1
CRYSTAL SYSTEM: Monoclinic. Usually in long prismatic crystals, occasionally granular. Prismatic crystals usually display sharp prisms.
OCCURRENCES: Crocoite is usually of secondary origin, derived from warm waters containing chromium acting on lead minerals, and is fairly rare. Good crystals are practically nonexistent in North America, but fair clusters have been obtained in the past in the Vulture District, Arizona, U.S. The best crystals with vivid colors are obtained from Dundas district, also at Heazlewood, and at White River, Tasmania, Australia. Crystals nearly always as crusts on limonite, or as clustered masses. Also found in Beresov District, U.S.S.R. Quite good crystal clusters at Goyabeira, Minas Gerais, Brazil.

Crocoite

NAME: **Wolframite**   FORMULA: $(Fe,Mn)WO_4$

PHYSICAL PROPERTIES:

Color: Dark grey or brownish black

Streak: Dark reddish brown   *Lustre:* Submetallic
   to black

Hardness: 5.0–5.5   *Sp. G.:* 7.1–7.5

CRYSTAL SYSTEM: Monoclinic. Crystals are usually thick tabular, but also prismatic. Sometimes bladed, columnar, or granular. Faces often striated.

OCCURRENCES: Wolframite is a high-temperature mineral of hydrothermal deposits, often with cassiterite, topaz, and scheelite. It is also found in pegmatites and granite. In U.S., fairly good crystals are found in the Quartz Creek district, Gunnison County, and at Silverton in San Juan County, Colorado. Some red-brown crystals occur at Townesville, North Carolina. Splendid crystals are present at Zinnwalt and Schlaggenwald, Czechoslovakia. Very large wedge-shaped crystals come from Llallagua, Bolivia.

Wolframite
(Dark grey-black crystals; associated with yellow Topaz crystals.)

NAME: **Scheelite**   FORMULA: $CaWO_4$

PHYSICAL PROPERTIES:

Color: White, pale yellow, brownish

Streak: White   *Lustre:* Vitreous to subadamantine

Hardness: 4.5–5.0   *Sp. G.:* 6.1

CRYSTAL SYSTEM: Tetragonal. Often in octahedral crystals, with pyramidal forms. Sometimes tabular. Also columnar, and massive granular.

OCCURRENCES: Scheelite is usually a high-temperature hydrothermal or contact-metasomatic mineral, occurring in veins or in tactite bodies. In U.S., nice orange-brown crystals up to 1″ in size are found near Milford, Utah. At the Boriana Mine and the Cohen Mine, Arizona, beautiful specimens are found. Also found as good crystals at Caldbeck Falls, Cumberland, England. Also in Saxony, Germany, and at Zinnwald, Czechoslovakia. Found also in Spain and Peru. Very fine small crystals occur at Traversella, Italy. Good 1″ specimens from Mina Perdida, Peru.

Scheelite

NAME: **Wulfenite**   FORMULA: $PbMoO_4$

PHYSICAL PROPERTIES:

Color: Orange-yellow to orange-brown; also tones of yellow

Streak: White   *Lustre:* Resinous to adamantine

Hardness: 2.5–3.0   *Sp. G.:* 6.5–7.0

CRYSTAL SYSTEM: Tetragonal. Usually in thin or thick tabular to blocky crystals. Often long prismatic with pyramidal forms. Frequently octahedral; also granular massive.

OCCURRENCES: Wulfenite is usually of secondary origin, being derived from lead and molybdenum minerals of hydrothermal deposits. In U.S., beautiful, bright orange-red tabular crystals found at Red Cloud, at Hamburg Mines, and at Old Yuma Mine, Arizona. Nice yellow crystals at Total Wreck Mine, at Mammoth Mine, and Sunrise Mine, Arizona. Also at Central District, New Mexico. Large splendid crystals at Villa Ahumada, Mexico. Found also at Mies, Yugoslavia, and at Pribram, Czechoslovakia. Large tabular crystals are recently reported from M'Foati, Congo.

114

1 ×

Wulfenite
(Orange-yellow crystals; on fine-grained rock.)

NAME: **Apatite**   FORMULA: $Ca_5(F,Cl,OH)(PO_4)_3$

PHYSICAL PROPERTIES:

Color: Sea green; often violet-blue; also white to brown

Streak: White   *Lustre:* Vitreous to subresinous

Hardness: 5.0   *Sp. G.:* 3.17–3.23

CRYSTAL SYSTEM: Hexagonal. Long to short prismatic crystals of good hexagonal cross-section. Also tabular: occasionally reniform; also fibrous or granular.

OCCURRENCES: Apatite is widely distributed in all types of rocks, but mostly in metamorphic rocks, such as marble, gneiss, and schist. In Canada, nice, large (2″–10″) crystals in Renfrew, Lamarck, Frontenac, and Haliburton Counties, Ontario, in marble. In U.S., good crystals are found at Pelham, Massachusetts. Very beautiful clear purplish crystals at Mt. Apatite, Maine; also in Himalaya Mine, California. Fine yellow crystals in open-pit mines, Ciudad Durango, Mexico. Also at Ehrenfriedersdort, Saxony, Germany; at Glesch, Switzerland. At Jumilla, Spain, small crystals on volcanic rock are present. Very excellent crystals, with tin, occur at Llallagua, Bolivia. Fine green crystals at Minas Gerais, Brazil.

Apatite

NAME: **Pyromorphite**   FORMULA: $Pb_5Cl(PO_4)_3$

PHYSICAL PROPERTIES:

Color: Green, yellow, brown

Streak: White to faintly tinted yellow   *Lustre:* Resinous

Hardness: 3.5–4.0   *Sp. G.:* 6.5–7.1

CRYSTAL SYSTEM: Hexagonal. Usually in prismatic crystals with good hexagonal cross-sections. Often globular to reniform; also fibrous.

OCCURRENCES: Pyromorphite is usually of secondary origin, produced in the oxidized portions of lead-bearing hydrothermal deposits. Very nice green and yellow small crystals are found at the Society Girl Mine at Moyie, British Columbia, Canada. In U.S., beautiful green prismatic crystals are present in the Coeur d'Alene district in Idaho, such as at the Hercules Mine and the Caledonia Mine. Good crystals occur at Ojuela Mine, Durango, Mexico. Also found at Bad Ems, Germany, as barrel-shaped brown crystals, and at Moses, Germany, as large brown crystals. Also at Pribram, Czechoslovakia; and at Broken Hill, New South Wales, Australia. Quite nice crystals occur in Cornwall, England.

Pyromorphite

NAME: **Vivianite**   FORMULA: $Fe_3(PO_4)_2.8H_2O$

PHYSICAL PROPERTIES:

Color: Colorless, blue-green, toward purplish

Streak: Colorless, bluish to   *Lustre:* Pearly to vitreous
  purplish

Hardness: 1.5–2.0   *Sp. G.:* 2.6

CRYSTAL SYSTEM: Monoclinic. Usually in slender prismatic or bladed crystals, arranged in radiating aggregates. Sometimes reniform and globular.

OCCURRENCES: Vivianite is usually a secondary mineral in sulfide vein deposits, but also occurs in clay and other sedimentary rocks associated with organic matter. In U.S., radiating clusters are found at Mullica Hill, at Allentown, and at Shrewsbury, New Jersey, mostly associated with fossils. Beautiful crystals also at Ibex mine at Leadville, Colorado. Blue, bladed crystals occur at Bingham, Utah. Also at other localities in western U.S. mining areas. Found in dark green crystals at Poopo and Llallagua, Bolivia. In the past, good crystals at St. Agnes and Truro, Cornwall, England. Very fine crystals to $1\frac{1}{2}''$ at Wannon River Falls, Victoria, Australia.

Vivianite
(Long purplish crystals.)

NAME: **Wavellite**   FORMULA: $Al_3(OH)_3(PO_4)_2 \cdot 5H_2O$
PHYSICAL PROPERTIES:
Color: White to yellow to green
Streak: White   *Lustre:* Pearly to vitreous
Hardness: 3.3–4.0   *Sp. G.:* 2.3
CRYSTAL SYSTEM: Orthorhombic. Well-shaped crystals are fairly rare, but acicular radiating clusters as crusts or spheres are common.
OCCURRENCES: Wavellite forms as a secondary mineral usually in the oxidized portion of veins, but in other rocks too. In U.S., very fine specimens are obtained in Arkansas from narrow openings in altered and brecciated novaculite. Near Pencil in Montgomery County, spheroidal aggregates up to 1″ or so in diameter are found. Aggregates of small crystals are found at Moores Mill, and at Hellerton, Pennsylvania. Also found at King Turquois Mine, Colorado. Found in microcrystals in tin veins at Llallagua, Bolivia. Also near Barnstaple, Devonshire, England, and in radiating masses at Back Creek, Tasmania, Australia.

Wavellite

NAME: **Turquoise**

FORMULA: $CuAl_6(OH)_8(PO_4)_4 \cdot 4H_2O$

PHYSICAL PROPERTIES:

Color: Sky blue, bluish green to apple green to fine blue

Streak: White to greenish white   *Lustre:* Slightly waxy

Hardness: 5.0–6.0   *Sp. G.:* 2.6–2.8

CRYSTAL SYSTEM: Triclinic. Crystals are rare. Turquoise is usually reniform, stalactitic, or massive. Also in thin seams and disseminated grains.

OCCURRENCES: This copper-phosphate mineral is usually of secondary origin, developed by surface waters in the upper portions of copper deposits. It usually takes on a reniform or nodular form. In U.S., it is common in copper deposits in Arizona, but is fairly rare in other parts of North America. Beautiful nodules are found at Battle Mountain, Nevada, but the largest quantities come from Arizona. Crystallized turquoise of small size is found at Lynch Station, Campbell County, Virginia. Also at Ali-Mirsa-Kuh Mountains near Nishapur, Iran. Fair to good material also is obtained at Chuquicamata, Chile. Pale material in Cornwall, England.

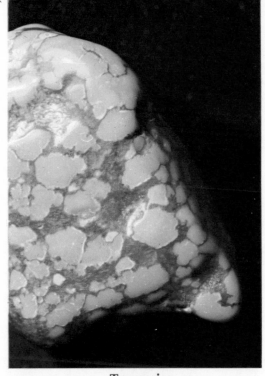

Turquoise
(Orange material is weathered rock.)

NAME: **Lazulite**   FORMULA: $(Mg,Fe)Al_2(OH)_2(PO_4)_2$
PHYSICAL PROPERTIES:
Color: Azure blue or deep blue
Streak: White   *Lustre:* Vitreous
Hardness: 5.0–6.0   *Sp. G.:* 3.1
CRYSTAL SYSTEM: Monoclinic. Crystals usually show acute pyramidal form. Often in veinlets and occasionally in granular aggregates.
OCCURRENCES: This blue mineral is found in highly metamorphosed quartzites, and also in pegmatites. In U.S., well-shaped crystals, displaying nice pyramids, are found at Graves Mountain, Lincoln County, Georgia. Deep blue masses up to 6″ across are present at the Champion sillimanite mine at Laws, California. Very fine crystals are found near Werfen, Salzburg, Austria. Also at Westana, Sweden. Gem crystals occur at Minas Gerais, Brazil.

126

Lazulite
(Blue crystals; with white Quartz.)

NAME: **Torbernite**

FORMULA: $Cu(UO_2)_2(PO_4)_2.8\text{-}12H_2O$

PHYSICAL PROPERTIES:

Color: Grass green to leek green

Streak: Pale green   *Lustre:* Pearly to adamantine

Hardness: 2.0–2.5   *Sp. G..* 3.2

CRYSTAL SYSTEM: Tetragonal. Crystals are usually square and tabular, commonly quite thin, occasionally displaying pyramidal faces.

OCCURRENCES: Torbernite usually occurs as a secondary mineral of oxidized uranium-bearing veins, probably derived most often from uraninite. In U.S., it occurs as small tabular crystals in pegmatites of New England, North Carolina, and the Black Hills of South Dakota. Very fine crystals of platy habit are found at Mina Candelaria near Moctezuma, Mexico. Found also as fine crystals at Gunnislake, near Callington, and at Redruth in Cornwall, England. Very splendid crystals at Mt. Painter, South Australia, and at South Alligator Gorge, Northern Territory, Australia. Also at Chinkolobwe, Congo.

Torbernite

NAME: **Autunite**

FORMULA: $Ca(UO_2)_2(PO_4)_2.10\text{-}12H_2O$

PHYSICAL PROPERTIES:

Color: Vivid lemon yellow to greenish yellow

Streak: Yellowish white   *Lustre:* Pearly to adamantine to yellowish green

Hardness: 2.0–2.5   *Sp. G.:* 3.1

CRYSTAL SYSTEM: Tetragonal. Usually in thin tabular crystals. Commonly in fan-like growths or in jumbled masses. Also foliated.

OCCURRENCES: This is a secondary mineral formed from uranium minerals, especially uraninite, in the oxidized zones of ore deposits. In U.S., well-shaped small crystals of autunite have been found in pegmatites of New England and North Carolina. Very fine crystals are tound at the Daybreak Mine, Spokane County, Washington. Beautiful specimens come from Mt. Painter and Katherine in Australia. Nice crystals also from Margac, France. Also obtained from Redruth and St. Austell, Cornwall, England.

Autunite

NAME: **Mimetite**   FORMULA: $Pb_5Cl(AsO_4)_3$

PHYSICAL PROPERTIES:

Color: Yellow, yellow-brown, brown

Streak: White with slight yellow tint   *Lustre:* Resinous

Hardness: 3.5–4.0   *Sp. G.:* 7.2

CRYSTAL SYSTEM: Hexagonal. Usually prismatic, with good hexagonal cross-sections. Often arranged in branching clusters. Also globular and fibrous.

OCCURRENCES: Mimetite is very similar to pyromorphite in composition, but is an arsenate, while pyromorphite is a phosphate. Mimetite is found usually with pyromorphite in the oxidized portions of sulfide deposits. In U.S., nice mimetite crystals of orange-yellow color are obtained at the 79 Mine in Arizona. Very splendid large crystals are found at Bilbao Mine, Mexico. Fine specimens at Wheal Alfred, in Cornwall, and at Dry Gill in Cumberland, England. At Dry Gill, specimens range from brown to bright yellow in color.

5 ×

Mimetite
(Large yellow crystals.)

NAME: **Erythrite**   FORMULA: $Co_3(AsO_4)_2.8H_2O$

PHYSICAL PROPERTIES:

Color: Crimson to peach red

Streak: Pale red to pink   *Lustre:* Vitreous to pearly

Hardness: 1.5–2.5   *Sp. G.:* 2.95

CRYSTAL SYSTEM: Monoclinic. Crystals are usually long prismatic and faces are usually striated. Also occurs as globular tufts of microcrystals, and often is dust-like and earthy.

OCCURRENCES: This mineral is usually developed as an alteration product of cobalt and nickel minerals, and often contains some nickel along with the cobalt. Crystals of poor quality are common in the Timiskaming District, Ontario, Canada, associated with nickel sulfide veins. Also found as fine microcrystals covering over cracks in rocks at Alamos, Mexico. Beautiful crystals were formerly obtained at Schneeberg, Germany, especially at Grube Rappold. Recently, splendid bladed crystals have been found at Bou Azzer, Morocco. The Moroccan specimens are in great demand and command high prices.

Erythrite

NAME: **Annabergite**   FORMULA: $Ni_3(AsO_4)_2.8H_2O$

PHYSICAL PROPERTIES:

Color: Fine apple green to yellow-green

Streak: Pale greenish white   *Lustre:* Vitreous to pearly

Hardness: 2.5–3.0   *Sp. G.:* 3.0

CRYSTAL SYSTEM: Monoclinic. Crystals usually in capillary types, usually appearing as crust or powders. Usually in microcrystals.

OCCURRENCES: Annabergite usually occurs as alteration products of nickel minerals and occasionally from cobalt minerals, as cobalt often is present along with the nickel. It is very similar to erythrite. Found along with erythrite in association with sulfide deposits in the Timiskaming District, Ontario, Canada. Recently found with erythrite at Bou Azzer, Morocco, in fine, bladed aggregates. Also was found at Laurium, Greece; in Saxony, Germany; and also at Chalantes, France. Annabergite is rare in the U.S.

Annabergite
(Light green crystals.)

NAME: **Pharmacolite**   FORMULA: $CaH(AsO_4).8H_2O$

PHYSICAL PROPERTIES:

Color: White or greyish white with slight red tinge

Streak: White   *Lustre:* Vitreous

Hardness: 2.0–2.5   *Sp. G.:* 2.7

CRYSTAL SYSTEM: Monoclinic. Usually as fine delicate silky fibres in radiating clusters. Also as botryoidal or stalactitic masses.

OCCURRENCES: Pharmacolite is a rare mineral, but it is an alteration product of arsenic-bearing minerals. It has been found at Joachimsthal, Czechoslovakia and at Andreasberg in the Harz Mountains. It occurs in fairly nice crystals in the Black Forest and at Glücksbrunn in Thuringia. At the Ste. Marie aux Mines in the Alsace region of France, pharmacolite occurs as botryoidal or globular masses.

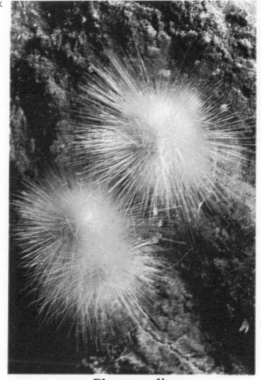

Pharmacolite

NAME: **Vanadinite**   FORMULA: $Pb_5Cl(VO_4)_3$

PHYSICAL PROPERTIES:

Color: Deep ruby red, reddish brown to brownish yellow

Streak: White to yellowish white   *Lustre:* Resinous

Hardness: 2.5–3.0   *Sp. G.:* 6.5–7.0

CRYSTAL SYSTEM: Hexagonal. Usually in prismatic crystals, often with small pyramidal forms. Often in parallel groups.

OCCURRENCES: Vanadinite is an uncommon mineral of secondary origin in oxidized portions of lead deposits. In U.S., very nice crystals are found at Fairview Claim, Black Canyon District, and from the Sierra Grande Mine in New Mexico. Found as especially fine brilliant red crystals at the Apache Mine and in the Globe district, Arizona. Good specimens occur at El Dorado Mine, California. Nice brown crystals come from Villa Ahumada, Mexico. Also at Leadhills district, Scotland. Very large handsome crystals occur at Djebel Mahseur, Morocco.

Vanadinite

NAME: **Descloizite**   FORMULA: $(Zn,Cu)Pb(OH)(VO_4)$
PHYSICAL PROPERTIES:
Color: Reddish brown to black; also cherry red
Streak: Orange to brownish red   *Lustre:* Greasy to grey
Hardness: 3.0–3.5   *Sp. G.:* 6.2
CRYSTAL SYSTEM: Orthorhombic. Usually in small drusy crystals or radiating fibres forming plumose growths. Good crystals usually display pyramidal forms.
OCCURRENCES: A secondary mineral in the oxidized zone of ore deposits containing lead and vanadium. In U.S., small crystals are found at many places in Arizona and New Mexico; also in Mexico. Good specimens have come from Pim Hill near Shrewsbury, England, occurring as crusts. At Otavi, South West Africa, large clusters have been found in the past. Other places in South West Africa where descloizite is found are Grootfontein, Friezenberg, Abenab, and Tsumeb.

Descloizite

NAME: **Garnet Group**

The Garnet Group includes a series of six varieties known collectively as the Garnets. They have very similar crystal structures, crystallize in the Isometric System, and commonly display dodecahedral and trapezohedral forms. Although they have similar formulas, the elements differ somewhat, and ionic substitution is common. The six theoretically pure species are described here:

OCCURRENCES: Garnet is usually found in metamorphic rocks in highly siliceous schists and gneisses. It also occurs in crystalline limestones, or sometimes in granites, and also in serpentine.

Almandite: FORMULA: $Fe_3Al_2(SiO_4)_3$; *Hardness:* 7.5; *Streak:* Colorless-White; *Sp. G.:* 3.9–4.3. Found in biotite schist along the Skeena and Stikine Rivers, British Columbia, Canada. In U.S., beautiful dodecahedrons of good size occur at Roxbury, Connecticut, and at Michigamme, Michigan. Also at Sedalia Mine, Salida, Colorado. Also at various places in India and in Africa.

Pyrope: FORMULA: $Mg_3Al_2(SiO_4)_3$; *Hardness:* 7.3; *Streak:* Colorless-White; *Sp. G.:* 3.6–3.8. Found in U.S. in Cowee Creek gravels in North Carolina. Pyrope is also associated with Arkansas diamonds, as well as at various places in Arizona, Utah, and New Mexico. Found also

Spessartite

4 X

Grossularite

5 ×

Andradite

at Trebnitz, Czechoslovakia. Also at Kimberly, South
Africa; in New South Wales, Australia.

Spessartite: FORMULA: $Mn_3Al_2(SiO_4)_3$; *Hardness*: 7.3;
Streak: Colorless-White; *Sp. G.*: 4.1–4.2. Crystals of
large size are present in U.S. in Rutherford Mine No. 2,
Amelia, Virginia, as well as in pegmatites of Ramona
District, California. Also at Pocos de Cavalos, Brazil and
from Madagascar.

Grossularite: FORMULA: $Ca_3Al_2(SiO_4)_3$; *Hardness*:
7.0; *Streak*: Colorless-White; *Sp. G.*: 3.6–3.7. Found in
Gatineau County and Megantic County, Quebec,
Canada. In U.S. nicely shaped crystals are present at
Minot, Maine; at Calumet Iron Mine, Colorado. Also
at Xalostoc, Mexico. Fine collecting spot is near
Piedmont, Italy.

Andradite: FORMULA: $Ca_3Fe_2(SiO_4)_3$; *Hardness*: 6.5;
Streak: Colorless-White; *Sp. G.*: 3.8–3.9. Found in U.S.
in zinc deposits of Franklin, New Jersey, with fine shapes
and good sizes. Also nearby at Sterling Hill, Ogdensburg.
Cornwall Iron Mine in Pennsylvania produces some
nicely striated crystals. Also at Walker, California. A
well-known locality is Sissersk District, Nizhni-Tagil,
U.S.S.R.

Uvarovite: FORMULA: $Ca_3Cr_2(SiO_4)_3$; *Hardness*: 7.5;
Streak: Colorless-White; *Sp. G.*: 3.4–3.5. A green garnet
rarely in good crystals, is found at Magog and at
Thetford Mines, Quebec, Canada. Also at Riddle,
Oregon, U.S. Large crystals at Outokumpu, Finland.

10 ×

Uvarovite

NAME: **Topaz** FORMULA: $Al_2(OH,F)SiO_4$

PHYSICAL PROPERTIES:

Color: Straw yellow, wine yellow, white, grey, greenish

Streak: Uncolored *Lustre:* Vitreous

Hardness: 8.0 *Sp. G.:* 3.4–3.6

CRYSTAL SYSTEM: Orthorhombic. Usually in prismatic crystals, with diamond-shaped cross-section. Prisms usually terminated by pyramids and basal pinacoids. Often granular.

OCCURRENCES: Topaz usually is associated with acidic igneous rocks, such as granites and rhyolites. Sometimes it is found in metamorphic rocks. In U.S., fine blue crystals in cavities in granite are present near Conway, New Hampshire. Many fine crystals are produced at Streeter, Mason County, Texas, in granitic rock of the Llano Uplift, both in pegmatites and stream gravels. Also in the Pikes Peak area of Colorado. Very fine crystals occur at Thomas Mountain, Juab County, Utah. Gem-quality topaz from Ouro Preto, Minas Gerais, Brazil. Some also in U.S.S.R., near Nerchinsk. Found in granite in Mourne Mountains, County Down, Ireland. In Australia, small but nice crystals come from Northern Territories and from Victoria.

Topaz

NAME: **Kyanite**   FORMULA: $Al_2O(SiO_4)$

PHYSICAL PROPERTIES:

Color: Blue to white; also green

Streak: Uncolored   *Lustre:* Vitreous to pearly

Hardness: 5.0–7.2   *Sp. G.:* 3.6

CRYSTAL SYSTEM: Triclinic. Long, bladed crystals are common, but terminating forms are unusual. Long, bladed aggregates diverge in sprays.

OCCURRENCES: This mineral usually is found in gneisses and mica schists. It occasionally is found in veins in metamorphic rocks. In U.S., blue bladed crystals up to 3″ long come from Baker Mountain, Virginia. Very splendid crystals are associated with rutile on Graves Mountain, Georgia. At Capelinha, Minas Gerais, Brazil, fine beautiful blue blades of good size have been found. A good collecting site is at Pizzo Forno, Switzerland, where pale blue crystals are found with staurolite. Found also in gravels in Ceylon; and beautiful bladed shapes in Tanganyika.

Kyanite
(Long blue crystals; in mica-schist rock.)

NAME: **Sphene** (Titanite)   FORMULA: $CaTiO(SiO_4)$
PHYSICAL PROPERTIES:
Color: Brown, yellow, green, or grey
Streak: Colorless or white *Lustre:* Adamantine to resinous
Hardness: 5.0–5.5   *Sp. G.:* 3.5
CRYSTAL SYSTEM: Monoclinic. Crystals usually wedge-shaped and flattened; but they usually display good prisms. Also massive or compact.
OCCURRENCES: Sphene usually is associated with granitic to intermediate igneous rocks, and also with metamorphic rocks. In the Grenville marble, in Ontario, Canada, large crystals are found associated with apatite and calcite, such as at Eganville. In U.S., very fine crystals up to 8″ across are also found at Rossie and Gouverneur, New York. A few sphene crystals have been found at Franklin, New Jersey. Gem quality 4″ sphere-like crystals occur at El Alamo in Baja California Norte, Mexico. Also found at numerous places in Switzerland, but rather hard to get.

Sphene
(Yellow crystal; lying on fine-grained
marble.)

NAME: **Datolite**   FORMULA: $Ca(OH)(BSiO_4)$

PHYSICAL PROPERTIES:

Color: White, sometimes pale green, red, or amethystine

Streak: White   *Lustre:* Vitreous

Hardness: 5.0–5.5   *Sp. G.:* 2.9–3.0

CRYSTAL SYSTEM: Monoclinic. Usually in short prismatic crystals with blunt wedge-shaped ends. Sometimes botryoidal, or columnar, or compact.

OCCURRENCES: Datolite is a secondary mineral in veins and cavities in basic volcanic rocks. Very fine crystals occur at the Smith-Lacey mine at Loughborough, Ontario, Canada. In U.S., extremely large crystals, up to 3″, occur in basalt fissures at Westfield, Massachusetts. Nice crystals are found at Rocky Hill and at East Granby, Connecticut. Found in large crystals, up to 2″ in size, at the Osceola Mine, as cavity-fillings in volcanics on the Keeweenaw Peninsula, Michigan. Beautiful crystals have been found in silver deposits at Guanajuato, Mexico. Also at Audreasburg, Germany; Trentino, Italy; and at Roseberry, Tasmania.

3 ✕

Datolite

NAME: **Uranophane**

FORMULA: $Ca(UO_2)_2(OH)_2(SiO_3)_2.5H_2O$

PHYSICAL PROPERTIES:

Color: Citrine to dark yellow

Streak: Light yellow   *Lustre:* Vitreous

Hardness: 2.5   *Sp. G.:* 3.8–3.9

CRYSTAL SYSTEM: Monoclinic. Usually in prismatic to fibrous radiating aggregates.

OCCURRENCES: Uranophane is usually a primary mineral in granites and other siliceous rocks. In U.S., it is found commonly in uranium deposits in New Mexico, and in the Black Hills of South Dakota. It is found also as an alteration mineral from gummite at mica mines of Mitchell County, North Carolina. Very fine crystals come from Menzenschwand, Schwarzwald, Germany. Also found in quite nicely shaped crystals at Wölsendorf, Bavaria.

Uranophane

NAME: **Hemimorphite**

FORMULA: $Zn_4(OH)_2(Si_2O_7) \cdot H_2O$

PHYSICAL PROPERTIES:

Color: White, with delicate bluish or greenish tone

Streak: White   *Lustre:* Vitreous

Hardness: 4.5–5.0   *Sp. G.:* 3.4–3.5

CRYSTAL SYSTEM: Orthorhombic. Crystals are usually tabular, but also prismatic. Crystals rarely doubly terminated, but show hemimorphic development. Commonly in sheaf-like masses; fan-shaped.

OCCURRENCES: Hemimorphite is found in the oxidized portions of zinc deposits. In U.S., a former famous collecting site was at Sterling Hill, New Jersey, where large, poorly formed crystalline masses were present. Reasonably fine crystals can be found at Elkhorn, Montana. Very nice white crystals are obtained at lead-zinc mines at Leadville, Colorado. Beautiful single blades of transparent crystals are found at Mina Ojuela, at Mapimi, Durango, Mexico. Good crystals at Nerchinsk, Siberia; at Altenberg, Germany; at Iglesias, Italy; and at Djebel Guergour, Algeria. Acicular crystals and crusts in Cumberland and near Matlock, Derbyshire, England.

Hemimorphite

NAME: **Epidote**

FORMULA: $Ca_2(Al,Fe)Al_2O\,(OH)(SiO_4)(Si_2O_7)$

PHYSICAL PROPERTIES:

Color: Pistachio green, yellowish green, brownish green

Streak: Uncolored to slightly greyish   *Lustre:* Vitreous

Hardness: 6.0–7.0   *Sp. G.:* 3.3–3.5

CRYSTAL SYSTEM: Monoclinic. Crystals are commonly prismatic, and are usually deeply striated. Often bladed aggregates. Also fibrous or granular.

OCCURRENCES: Epidote is commonly found in low-grade metamorphic rocks, but also in contact areas of metamorphic marbles. In U.S., very fine short stubby prismatic crystals up to 1″ in size are found at the Calumet Iron Mine and at Epidote Hill, Colorado. Extremely large, 12″ crystals at the Peacock Mine and at the Decorah Mine in the Seven Devils district, Idaho. Also in Greenhorn Mountains, and at Riverside, California. Found at Arendal, Norway, and at the Knappenwand, near Salzburg, Austria. A good collecting site is near Sulzer, Prince of Wales Island, southwestern Alaska.

Epidote
(Long green crystals; white fibrous crystals
are Tremolite.)

NAME: **Vesuvianite** (Idocrase)

FORMULA: $Ca_{10}Mg_2Al_4(OH)_4(Si_2O_7)_2(SiO_4)_5$

PHYSICAL PROPERTIES:

Color: Brown to green

Streak: White　　　　　　　　　　*Lustre:* Vitreous

Hardness: 6.5　　　　　　　　　　*Sp. G.:* 3.4

CRYSTAL SYSTEM: Tetragonal. Commonly in stubby prismatic crystals displaying pyramids. Also massive, columnar.

OCCURRENCES: Vesuvianite is usually found in contact metamorphic limestone masses, but it is somewhat rare. Also it is found in pegmatites. In Canada very fine, yellow gem crystals are found in pegmatite near Laurel, Quebec, and good pink crystals are found at Black Lake, Quebec. Very nice crystals occur in marble at Marble Bay, Texada Island, British Columbia. In U.S., brownish green crystals occur at Sanford, Maine. Many other places in U.S. Good 2″ crystals at Kristiansand and Eiker, Oslo District, Norway. Good pyramidal crystals come from the vicinity of Mt. Vesuvius, Italy. Stubby green crystals are found at Achmatovsk, Siberia.

2 ×

Vesuvianite (Idocrase)

NAME: **Prehnite**   FORMULA: $Ca_2Al_2(OH)_2(Si_3O_{10})$
PHYSICAL PROPERTIES:
Color: Pale green to white
Streak: Colorless   *Lustre:* Vitreous to pearly
Hardness: 6.5   *Sp. G.:* 2.8–2.9
CRYSTAL SYSTEM: Orthorhombic. Crystals usually tabular, but also prismatic; rarely in single isolated crystals. Commonly in groups. Also stalactitic.
OCCURRENCES: Prehnite usually occurs in cavities in basalts having formed as secondary minerals. Occasionally found in cavities in granites. In U.S., very fine, large (12″) crystals are found at New Street Quarry and Prospect Park Quarry, Paterson, New Jersey. Splendid crystals associated with apophyllite from quarries in Fairfax and London Counties, Virginia. Quite good specimens occur in Keeweenaw Peninsula, Michigan, in copper deposits. Also at Motta Naira, Switzerland. In basalts at Renfrewshire and Dumbartonshire, Scotland. Found in yellow masses at Prospect, New South Wales.

2×

Prehnite

NAME: **Axinite**

FORMULA: $Ca_2(Fe,Mg,Mn)Al_2(BO_3)(OH)(Si_4O_{12})$

PHYSICAL PROPERTIES:

Color: Clove brown to blue to grey

Streak: Uncolored   *Lustre:* Glassy to vitreous

Hardness: 6.5–7.0   *Sp. G.:* 3.3

CRYSTAL SYSTEM: Triclinic. Usually broad crystals, with knife-like edges. Also in bladed aggregates. Occasionally granular.

OCCURRENCES: Axinite is associated with high-temperature hydrothermal deposits; also in contact metamorphic limestones, as well as in granitic rocks. In U.S., nice small, yellowish, drusy crystals have been found at Franklin, New Jersey, associated with zinc ores. Large bladed crystals are obtained near Luning, Nevada. Exceptional crystals have come from Coarse Gold, California. Splendid 2″ crystals come from scheelite mines at Los Gavilanes, Mexico. Probably the best crystals are from St. Christophe, near Bourg d'Oisans, France. Very fine specimens have been found in Cornwall, England; also at Toroku Mine, Kyushu Island, Japan.

Axinite

2 ×

Beryl
(Green crystals; in rock.)

NAME: **Beryl**   FORMULA: $Al_2Be_3(Si_6O_{18})$

PHYSICAL PROPERTIES:
Color: Greenish yellow, green, or greenish blue
Streak: Colorless   *Lustre:* Vitreous to slightly silky
Hardness: 8.0   *Sp. G.:* 2.7–2.9

CRYSTAL SYSTEM: Hexagonal. Usually occurs in long prismatic crystals, terminating in pinacoids and pyramids. Often coarse columnar to granular, also.

OCCURRENCES: Beryl is most frequently associated in granitic pegmatites, but occasionally in granites. Blue beryl occurs in pegmatites in Ontario, Canada, particularly at Lyndoch. In U.S., fairly abundant in

170

Beryl

pegmatites of New England, sometimes in very large crystals. Large 8″ emerald crystals have been found in North Carolina pegmatites at Hiddenite and at Shelby. Aquamarine is found in pegmatites in Colorado, California, and South Dakota. Good gem beryl comes from Columbia and from Minas Gerais, Brazil, usually as outstanding green varieties. Also found in Ireland, U.S.S.R., and in Madagascar.

171

Tourmaline (Rubellite)
(Pink crystals in rock.)

2 ×

NAME: **Tourmaline**

FORMULA: $Na(Mg,Fe)_3Al_6(OH)_4(BO_3)_3(Si_6O_{18})$

PHYSICAL PROPERTIES:

Color: Black commonly, but also green, pink, blue, yellow

Streak: Colorless (or white)   *Lustre:* Vitreous to silky

Hardness: 7.5   *Sp. G.:* 3.0–3.3

CRYSTAL SYSTEM: Hexagonal. Usually in long or short prismatic crystals, terminated by low pyramids. Commonly in six-sided or nearly triangular shaped crystals, displaying radiating aggregates.

OCCURRENCES: Commonly found in pegmatites, in

172

1 ×   **Tourmaline** (Verdelite)
(Green crystals; with white Quartz.)

metamorphic rocks, and in hydrothermal deposits.
Many localities exist where tourmaline is found. In U.S.,
nice small crystals are present at Pierrepont, New York.
Very fine red crystals are at Mount Mica, Maine.
Excellent green tourmalines are obtained from the
Gillette Quarry, near Haddam Neck, Connecticut.
Good gem crystals up to 6″ come from Himalaya Mine at
Mesa Grande, California. Also found at Alamos, Baja
California Norte, Mexico. Found in Urals in Russia;
also in Austria; in Madagascar; in South West Africa;
in Western Australia and South Australia. Good gem
material comes from Minas Gerais, Brazil.

173

NAME: **Chrysocolla**   FORMULA: $CuSiO_3.2H_2O$

PHYSICAL PROPERTIES:

Color: Green, bluish green, sky blue, blue
Streak: White to pale bluish green   *Lustre:* Vitreous
Hardness: 2.4   *Sp. G.:* 2.2

CRYSTAL SYSTEM: Cryptocrystalline, often like opal, as having formed from a gel. Sometimes earthy, or rounded masses.

OCCURRENCES: Chrysocolla is usually of secondary origin, associated with other copper minerals in hydrothermal deposits. In U.S., quite nice bluish green crystals come from Bisbee, Arizona, and have in the past come from Clifton-Morenci, from Globe district, and other Arizona deposits. Chrysocolla has come from copper mines of Cŏrnwall and Cumberland, England. Quite fine specimens are found at Katanga, Belgian Congo, and from copper deposits of Chile. Chrysocolla masses have come from the Ural Mountains of U.S.S.R.

2

Chrysocolla

NAME: **Augite**   FORMULA: $CaMgSi_2O_6$

PHYSICAL PROPERTIES:

Color: Black

Streak: Colorless   *Lustre:* Vitreous

Hardness: 6.0   *Sp. G.:* 3.2–3.5

CRYSTAL SYSTEM: Monoclinic. Crystals are usually short prismatic, terminated by pinacoids appearing to form wedges. Occasionally in long prismatic crystals.

OCCURRENCES: Augite is fairly common in the dark igneous rocks such as basalts and gabbros. In U.S., very nice augite crystals, up to $\frac{1}{2}''$ in size, are found in basalt at the Trail Creek-Gold Run Creek area of Grand County, Colorado. In Oregon, fairly good sized greenish-black crystals of augite are found in basalt at Cedar Brook near Tillamook. Very fine augites are found in basalt at Bufaure, Italy; and at Mt. Vesuvius, and at Mt. Etna, Sicily. Good augite crystals are found in basalt at Boreslau, Czechoslovakia.

3

Augite
(Black crystal.)

NAME: **Tremolite**   FORMULA: $Ca_2Mg_5(OH)_2(Si_4O_{11})_2$

PHYSICAL PROPERTIES:

Color: White or grey, or pale green

Streak: White to grey   *Lustre:* Vitreous

Hardness: 5.5–6.0   *Sp. G.:* 2.9–3.2

CRYSTAL SYSTEM: Monoclinic. Usually in distinct, fairly long-bladed crystals, although short stout crystals are fairly common. Often in long thin radiating aggregates, or fibrous.

OCCURRENCES: Tremolite is usually associated with metamorphic limestones and in schists. Very fine crystals are located in the Grenville marbles of Ontario, Canada, for example, at Haliburton and Wilberforce. In U.S., good pink crystals are found at DeKalb and at Edwards in New York. When the iron content increases, tremolite takes on a green color, such as at Pelham, Massachusetts. Good crystals are found near Jade Mountain along the Keokuk River in Alaska. Also found in Wyoming. Very finely crystallized white tremolite occurs at Passo de Campolungo, Tessin, Switzerland, and at other Swiss localities.

3×

Tremolite

NAME: **Actinolite**

FORMULA: $Ca_2(Mg,Fe)_5(OH)_2(Si_4O_{11})_2$ (similar to tremolite)

PHYSICAL PROPERTIES:

Color: Bright green and greyish green. Tremolite and Actinolite generally considered nearly identical except for more Fe in Actinolite. Thus white in Tremolite becomes green with increasing iron (Fe) content in Actinolite.

Streak: Colorless   *Lustre:* Vitreous

Hardness: 5.0–6.0   *Sp. G.:* 3.0–3.3

CRYSTAL SYSTEM: Monoclinic. Crystals are usually long and acicular, either in straight fibrous masses (asbestos) or in felted fibrous aggregates.

OCCURRENCES: Actinolite is usually of metamorphic origin, and is found in marbles and schists. In U.S., nice green crystals up to 3″ long are in asbestos masses at Pelham, Massachusetts, and in 5″ blades in talc near Chester, Vermont. Good crystals occur at Wrightwood, in San Bernardino County, California. Very fine long, acicular crystals occur in several Swiss areas.

3 ×

Actinolite
(Long green crystals; in metamorphosed dolomite rock.)

181

NAME: **Pectolite**   FORMULA: $Ca_4Na_2(Si_3O_9)_2 \cdot H_2O$

PHYSICAL PROPERTIES:

Color: White to greyish white, often pinkish white

Streak: Colorless   *Lustre:* Silky to vitreous

Hardness: 5.0   *Sp. G.:* 2.7–2.9

CRYSTAL SYSTEM: Triclinic. Crystals usually thin and acicular; commonly in compact radiating masses, and appear as globular shapes.

OCCURRENCES: Pectolite is usually of secondary origin, forming in cavities in volcanic rocks. Also, it forms in metamorphic rocks occasionally. The world's best specimens occur in U.S. in pillow basalt cavities in the northern part of New Jersey, such as at the Prospect Park quarry at Paterson. Also from Bergen Hill in Hudson County. Clusters of pectolite up to 8″ or 10″ in size have been found. Fairly coarse crystals have been found at Magnet Cove, Arkansas (as shown in foreground on page 183). Also present near Middletown and at Elder Creek, California. Formerly from the Val di Fassa, Italy; from Weardale, Durham, England, and from Ratho, Scotland.

Pectolite
(Both compact radiating mass of crystals
and large pink crystal.)

NAME: **Rhodonite**   FORMULA: $MnSiO_3$

PHYSICAL PROPERTIES:

Color: Rose pink to brownish red

Streak: White   *Lustre:* Vitreous

Hardness: 6.0   *Sp. G.:* 3.6–3.7

CRYSTAL SYSTEM: Triclinic. Crystals are usually tabular, but rather thick and appear blocky. Often in large crystals with nearly rounded ends, displaying prisms.

OCCURRENCES: Rhodonite may form during the metamorphism of manganese ores, or by contact metasomatic processes, or by hydrothermal processes. Good crystals are pretty rare. In U.S., very fine crystals of large tabular shapes were found at Franklin, New Jersey. Although this area is well picked over, rhodonite crystals can still be found. Small crystals are found at Pajsberg, Sweden. Blocky prismatic crystals up to 1″ in size are found with galena at Broken Hill, New South Wales, Australia. It is found also at Ekaterinberg, in the Ural Mountains, Russia.

Rhodonite

NAME: **Apophyllite** FORMULA: $KCa_4F(Si_4O_{10})_2 \cdot 8H_2O$

PHYSICAL PROPERTIES:

Color: White, often with grey or green tint; also rose.

Streak: White   *Lustre:* Pearly to vitreous

Hardness: 4.5–5.0   *Sp. G.:* 2.4

CRYSTAL SYSTEM: Tetragonal. Usually in fine square prisms terminated by pinacoids and pyramids. Sometimes pyramids are very prominent and prisms and pinacoids may be severely reduced in size.

OCCURRENCES: Commonly found in cavities in basaltic rock, or in granites and gneisses, and also in hydrothermal deposits. In Canada, good crystals of apophyllite have been found at Cape d'Or, Nova Scotia, and in the Rossland area of British Columbia. In U.S., very fine 1″ to 2″ crystals are present in the basalts near Paterson, at Bergen Hill, and at Snake Hill in New Jersey. Splendid 3″ tabular crystals are found near Centreville, Virginia, in diabase rock. Large pale-green crystals of apophyllite are found at Rio Grande Do Sul, Brazil. The best locality of former times was in the basalts of Poona, India. Also found at Aussig, Czechoslovakia; and Teigarhorn, Iceland.

Apophyllite
(White to light grey crystals in cavity in grey
basalt rock.)

NAME: **Serpentine**   FORMULA: $Mg_6(OH)_8(Si_4O_{10})$

PHYSICAL PROPERTIES:

Color: Olive green or blackish green

Streak: Colorless   *Lustre:* Waxy when massive and silky when fibrous

Hardness: Somewhat variable between 2.5 and 5.0   *Sp. G.:* 2.5–2.6

CRYSTAL SYSTEM: Monoclinic. Two common types: massive and fibrous. However, good definite individual crystals are rare. Common massive varieties are very finely crystalline in structure and appear nearly crypto-crystalline. Fibrous types (asbestos) are in prominent long fibres.

OCCURRENCES: Commonly associated with metamorphic rocks, being formed by the alteration of other magnesium-bearing minerals, such as olivine and enstatite. Serpentine is found as relatively small masses in schists and marbles. In U.S., translucent gem-quality serpentine is found near Rock Springs, Pennsylvania, where it occurs with chromite. Serpentine is found also at Montville, New Jersey, and in Gila County, Arizona. Very fine specimens of serpentine also occur at Snarum, Norway. A very well known English locality is the Lizard peninsula in Cornwall, where massive type of serpentine is cut and polished for ornamental purposes. Good quality ornamental serpentine is found also in Greece and Italy. Excellent specimens of fibrous serpentine (asbestos) come from the Thetford Mines area in Quebec, Canada,

188

Serpentine

where masses several inches thick and fibres up to 7″ are obtained.

NAME: **Mica** (The group of micas consist of a series of minerals which have very similar structures and compositions, but differ in color).

FORMULA: Simplest formula: $KAl_2(OH)_2(AlSi_3O_{10})$

PHYSICAL PROPERTIES:

Color: **Muscovite:** silver white, colorless, grey, and usually shining; **Biotite:** black, dark brown, brownish green; **Phlogopite:** yellowish brown to reddish brown; **Zinnwaldite:** brown to pale violet; **Lepidolite:** pale lilac to rose.

Streak: Colorless   *Lustre:* Pearly to vitreous

Hardness: 2.5   *Sp. G.:* 2.7–3.3

CRYSTAL SYSTEM: Monoclinic. Well-shaped crystals are uncommon, but usually are pseudo-hexagonal. Large, flat, shining surfaces are common; these are pinacoids. Usually in thin plates, blades, or scaly aggregates.

OCCURRENCES: The micas are usually found in granitic rocks, particularly in pegmatites. Often found in metamorphic rocks as well. Good **muscovite** crystals are found at the Rutherford Mines at Amelia, Virginia, and at several places in North Carolina, U.S. Good **biotite** crystals are rather scarce, but have been found at Mount Somma, Vesuvius, Italy. Very large **phlogopite** crystals come from the Lacy Mine in Ontario, Canada, and in U.S. from St. Lawrence County, New York, and from Sussex County, New Jersey. Fine crystals of **lepidolite** are found in pegmatites in the U.S. in Black Hills, Wyoming, and in California, particularly

190

Phlogopite

at Pala, San Diego County. **Zinnwaldite** is fairly rare, but has been found in Cornwall, England, and in the York district of Alaska.

NAME: **Quartz** FORMULA: SiO_2

PHYSICAL PROPERTIES:

Color: Colorless when pure; often yellow, pink, red, brown, green

Streak: White *Lustre:* Vitreous

Hardness: 7.0 *Sp. G.:* 2.65

CRYSTAL SYSTEM: Hexagonal. Often in nice prismatic crystals terminated by rhombohedrons. Often massive; also flint-like, mammillary, and concretionary. Quartz occurs in two main types, macrocrystalline and cryptocrystalline, and there are several varieties of each type.

Common quartz is usually colorless to white to nearly opaque, often occurring in large masses in veins and pegmatites. Good crystals are usually not developed, but short to long prismatic crystals are found.

Rock crystal: Colorless, in well-formed long prismatic crystals, often in drusy or radiating clusters; occasionally fibrous. Sharp, prismatic crystals are found at Black Rapids, Ontario, Canada. In U.S., in pegmatites in New England and North Carolina. Also at Mokelumne Hill, Calaveras County, California. Fine crystals in Herkimer and St. Lawrence Counties, New York. Also at Hot Springs, Arkansas. Other localities include Madagascar, Minas Gerais, Brazil, and Carrara, Italy. Many localities in Swiss Alps. In England, fine crystals come from Frizington and other places in Cumberland.

Smoky quartz: Smoky yellow to dark smoky brown, varying to brownish black. Nice crystals of smoky quartz

192

× 1/2

Quartz-Rock Crystal

are found in basalts in Nova Scotia, Canada; in U.S., in pegmatites of New England and North Carolina. Fine specimens occur at Pikes Peak, Colorado. Also at Butte, Montana. Some found in the Cairngorm Mountains of Scotland. Many nice specimens are found at Grimsel, Switzerland, and many other Swiss locations. Also at Porretta, Italy; Madagascar; at Otomezaka, Japan.

Amethyst: Pale to dark violet or purple; color caused by impurity, probably manganese. Good amethyst crystals in pegmatites in Nova Scotia, Canada, and North Carolina, in U.S. Also on Keeweenaw Peninsula, Michigan. Excellent crystals with hematite up to 5″ in size at Thunder Bay, Ontario, Canada. Good crystals from cavities in basalts in New England states and Pennsylvania. Beautiful long tapering amethyst found at Guerrero and drusy crystals associated with silver at Guanajuato, Mexico. Also at Zillertal, Austria. Famous locality is at Mursinska, U.S.S.R.

Rose quartz: Pale to medium pink, but not usually in good crystals. Often massive. Color probably due to titanium. Found most often in pegmatites in massive forms. In U.S., common in pegmatites in North Carolina, in Black Hills, and in New England states. Large masses of tiny crystals on milky quartz found at Galilea, and very fine crystals at Governador Valadares, Minas Gerais, Brazil. Many in Switzerland and other European mining districts.

Smoky Quartz

Quartz-Amethyst

Rose Quartz

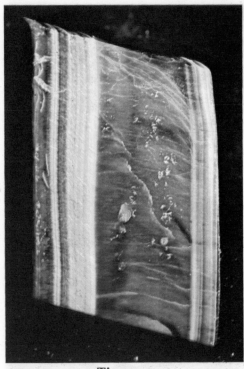

Tiger-eye

× 1/2 **Agate**

Tiger-eye: Yellow, brown, or bluish colored quartz which is pseudomorphous after crocidolite asbestos. Retains the fibrous structure of the asbestos. Tiger-eye is found near Griquatown, South Africa.

Agate: A variegated chalcedony in which colors are banded or clouded. Colors range from white to pale or dark brown to blues and other tones, usually in zigzag pattern, but approximately concentric circular. Beautiful agates are quite widespread, occurring in California, Mexico, Uruguay, Brazil, and various places in Europe.

199

Moss Agate

Moss Agate: White or cream-colored chalcedony containing wispy brown or black dendritic moss-like accumulations of manganese oxides, chlorite, or iron oxides. Beautiful moss agate has been found in Uruguay.
Jasper: Opaque chalcedony. Usually red, yellow, dark green, and greyish blue, depending upon the quantity of impurities such as clay, iron oxides, etc. Beautifully colored masses of jasper are used in the jewelry or ornament business. Heliotrope-colored jasper is found on the Kathiawar Peninsula, India.

200

Jasper
(Rare blue specimen.)

NAME: **Opal**   FORMULA: $SiO_2 . nH_2O$

PHYSICAL PROPERTIES:

Color: Colorless, sometimes white, also grey, brown or red

Streak: White   *Lustre:* Vitreous to slightly waxy

Hardness: 5.5–6.5   *Sp. G.:* 1.95–2.25

CRYSTAL SYSTEM: No crystal structure. Opal is amorphous. Usually massive, often in rounded forms; also stalactitic.

OCCURRENCES: Opal is found in low-temperature hydrothermal deposits. Also as nodules, seams, and crusts in a variety of environments. In U.S., thin films of fluorescent opal are found in pegmatites in New England and North Carolina. Large beautiful "wood opal" logs are found in Washington, Oregon, Idaho, Utah, and Nevada, especially at Virgin Valley, Humboldt County, Nevada. Excellent precious opal from volcanic rocks in Mexico; also in Honduras; and in Brazil. The finest opals are at the Lightning Ridge and White Cliffs areas of New South Wales, Australia. Prized as a gemstone.

Opal

Feldspars

The feldspars form a series of minerals which are closely related in composition and crystal structure. The feldspars are very abundant. They are classified into two groups: the **potassium feldspars** and the **sodium-calcium plagioclase feldspars.** The potassium feldspars include orthoclase, sanidine, and microcline, each bearing the same formula $KAlSi_3O_8$, but very slightly different crystal structure. The plagioclase feldspars form a continuous series from sodium-rich albite to calcium-rich anorthite with the general formula $(Na, Ca)(AlSi)(AlSi_2O_8)$. The similar crystal structure among the feldspars leads to the development of blocky crystals, with cleavages nearly at right angles, similar hardnesses, and other similar physical properties.

NAME: **Orthoclase**   FORMULA: $KAlSi_3O_8$

PHYSICAL PROPERTIES:

Color: White to pink to brownish pink to flesh colored
Streak: White   *Lustre:* Vitreous
Hardness: 6.0   *Sp. G.:* 2.5–2.6

CRYSTAL SYSTEM: Monoclinic. Crystals usually prismatic and blocky, but also thin tabular. Crystals usually display good prisms and pinacoids. Often massive. Twinned crystals are very common.

OCCURRENCES: Orthoclase is commonly found in granites and other kindred igneous rocks, often as single or twinned crystals. Good blocky orthoclase crystals are

Orthoclase × 1/2

found in 2″ sizes in granites in Penticton, British Columbia, Canada. In U.S., very fine 3″ crystals are present in quartz monzonite in the Tenmile district in Colorado; also in Gunnison County, Colorado. Excellent 3″ orthoclase crystals are found in the Sandia Mountains and in the Burro Mountains of New Mexico. Fine crystals in granite at Baveno quarries, Italy; also at Karlovy Vary, Czechoslovakia. Also found at Itrongahy, Madagascar. Crystal from Omi and Mino, Japan, are excellent.

NAME: **Adularia** (a variety of orthoclase)

FORMULA: $KAlSi_3O_8$

PHYSICAL PROPERTIES:

Color: White to light tan

Streak: White

Hardness: 6.0

Lustre: Vitreous

Sp. G.: 2.5–2.6

CRYSTAL SYSTEM: Monoclinic. Crystals are usually wedge-shaped or pseudorhombohedral.

OCCURRENCES: Adularia usually forms in low-temperature hydrothermal deposits. In U.S., very fine 4″ white adularia crystals are present at Sulzer, Prince of Wales Island, Alaska. Adularia crystals of fine white color are found in mines of the Silver City district, Idaho. Good large crystals are found at many places in Switzerland, as well as in Northern Italy and western Austria. The crystals from Switzerland and the nearby Alps are much in demand because of their large size and associated minerals, making striking contrasting specimens.

Adularia

NAME: **Microcline** (Amazonite)   FORMULA: $KAlSi_3O_8$
PHYSICAL PROPERTIES:
Color: Greenish or bluish green, also white
Streak: White   *Lustre:* Vitreous
Hardness: 6.0–6.5   *Sp. G.:* 2.5–2.6
CRYSTAL SYSTEM: Triclinic. Crystals are usually prismatic and blocky, but also tabular. Crystals usually display good prisms and pinacoids, and are often twinned.
OCCURRENCES: Microcline is primarily found in granitic pegmatites. In U.S., reasonably good crystals are found in pegmatites in Maine, New Hampshire, and North Carolina. Very fine ones are found in pegmatites of the Pikes Peak area and adjacent Teller County, Colorado. Some of the Pikes Peak amazonite were deep blue-green and up to 16″ in size. Gem-quality amazonite has been found in the Rutherford No. 1 and Morefield mines at Amelia, Virginia. Also found in Lyndock Twp, Renfrew County, Ontario, Canada. Other localities include Brazil, U.S.S.R., and Japan.

Microcline
(Pale green to white crystals.)

NAME: **Albite**   FORMULA: $NaAlSi_3O_8$

PHYSICAL PROPERTIES:

Color: White, sometimes faintly tinted

Streak: Colorless   *Lustre:* Vitreous

Hardness: 6.0–6.5   *Sp. G.:* 2.6

CRYSTAL SYSTEM: Triclinic. Crystals are often tabular or prismatic, showing good prisms and pinacoids. Sometimes occurs in massive or granular form. Twinning is very common, often as thin lamellar type.

OCCURRENCES: Albite is found usually in granitic pegmatites as bladed or tabular crystals in cavities. Some of the world's best albite crystals, up to 4″ in size, come from the Rutherford Mines at Amelia, Virginia, U.S. Very fine crystals are found in pegmatites of San Diego County, California, as well as in various places in Colorado. In Brazil, good albite has been found in pegmatites of Minas Gerais. It is also present in many localities in the Alpine regions, such as at St. Gotthard; at Alp Ruschuna near Vals; at Bristenstock near Amsteg; at Schmirn, Austria.

Albite

Zeolites

The zeolites, as a group, show close similarities in mode of occurrence, association, and composition. They are all hydrous silicates of aluminium, chiefly with either sodium or calcium. In gross composition, they resemble the feldspars. Zeolites most commonly are found in cavities and veins in basic igneous rocks. An interesting feature is that they have an open atomic structure with wide channelways where water molecules are held. Simple heating causes the water molecules to be given off without collapsing the structure.

NAME: **Analcite** (Analcime)

FORMULA: $Na(AlSi_2O_6).H_2O$

PHYSICAL PROPERTIES:

Color: Colorless, white, grey, yellow, reddish white

Streak: White   *Lustre:* Vitreous

Hardness: 5.0–5.5   *Sp. G.:* 2.3

CRYSTAL SYSTEM: Isometric. Crystals usually exhibit good trapezohedral faces, with or without cube faces. Often forms granular masses.

OCCURRENCES: Analcite is commonly found in cavities in basalts as a secondary mineral, but also occurs as a primary mineral in certain basic igneous rocks. Very fine brilliant crystals up to 4″ in size are recorded at Cape Blomidon, Nova Scotia, Canada. In U.S., excellent specimens, sometimes red, occur near Paterson, New Jersey. Splendid crystals come from Table Mountain,

Analcite
(Large orange crystal; with yellow crystals of
other Zeolites.)

Colorado, and also from Point Sal, California. Very
large pink crystals are found in basalt cavities at Seisser
Alp, Italy, and also from Sicily. Also from Aussig,
Czechoslovakia, and from Germany. Found on the Isle
of Skye, Scotland, from Prospect, New South Wales, and
from Flinders Island, Tasmania.

NAME: **Natrolite**   FORMULA: $Na_2Al_2Si_3O_{10}.2H_2O$

PHYSICAL PROPERTIES:

Color: White to colorless

Streak: Colorless   *Lustre:* Vitreous

Hardness: 5.5–6.0   *Sp. G.:* 2.25

CRYSTAL SYSTEM: Orthorhombic. Crystals are usually long prismatic, slender, or acicular; often fibrous, radiating aggregates. Good prismatic crystals display prisms and pinacoids.

OCCURRENCES: Natrolite often is found in cavities of basalts, as well as being a product of alteration of other minerals. In U.S., it is found as short prismatic crystals of small size ($\frac{1}{16}''$) at Snake Hill, Bergen County, and at the Houdaille Quarry in Somerset County, New Jersey. Very fine crystals occur at Table Mountain, Colorado. Also at Medford, Oregon. Large crystals up to $\frac{1}{2}''$ thick occur at Clear Creek, San Benito County, California. Additional localities include Aussig, Czechoslovakia; several places in the Alps; in France, particularly in the Auvergne. Very colorful orange-white specimens come from Hohentwiel, Germany.

2

Natrolite

NAME: **Stilbite** (Desmine)

FORMULA: $CaAl_2Si_7O_{18}.7H_2O$

PHYSICAL PROPERTIES:

Color: White to pale brown to dark brown

Streak: Colorless   *Lustre:* Pearly or vitreous

Hardness: 4.0   *Sp. G.:* 2.1

CRYSTAL SYSTEM: Monoclinic. Crystals are usually thin tabular, and are often grouped in nearly parallel position, forming sheaf-like bundles. Often in twinned form, but not easily observed.

OCCURRENCES: Stilbite is usually of secondary origin and is often found in cavities in basalt. It often is present in pegmatites, also. In U.S., some very nice crystals are found in pegmatites of the Pala area, San Diego County, California. Good specimens up to 3″ in size, also are found at Paterson, New Jersey. Other U.S. localities include Mitchell, Edwards, and Gable, Oregon. Basalts at Teigarhorn, Iceland, contain nice white bladed crystals. Cape d'Or in Nova Scotia, Canada, yields good specimens. It is present in silver veins at Guanajuato, Mexico. Also at Rio Grande Do Sul, Brazil. Found also at Kilpatrick, Scotland; Aussig, Czechoslovakia; Mt. Painter, Southern Australia.

3

Stilbite (Desmine)

NAME: **Harmotome**   FORMULA: $Ba(Al_2Si_6O_{16}).6H_2O$

PHYSICAL PROPERTIES:

Color: White to slightly grey to slightly pinkish

Streak: White   *Lustre:* Vitreous

Hardness: 4.5   *Sp. G.:* 2.5

CRYSTAL SYSTEM: Monoclinic. Crystals very commonly occur as interpenetration twins, each of which displays good prisms and oftentimes pinacoids.

OCCURRENCES: Harmotome is commonly found in cavities in basalts and other volcanic rocks. But it also has been reported in gneiss and in hydrothermal deposits. This mineral is not common in the U.S., but it has been found in pretty fair crystals in diabase near Ossining, New York. Nice cruciform twins up to $\frac{1}{2}''$ in size are found at the Korsnas Mine, Finland. Also found at Kongsberg, Norway. Very good crystals have come from the Bells Grove Mine, Strontian, Argyllshire, in Scotland. Another good source is near Andreasburg, Harz, Germany.

Harmotome

NAME: **Chabazite**   FORMULA: $(CaNa)Al_2Si_4O_{12}.6H_2O$

PHYSICAL PROPERTIES:

Color: White to flesh red

Streak: Colorless to white   *Lustre:* Vitreous

Hardness: 4.5   *Sp. G.:* 2.1

CRYSTAL SYSTEM: Hexagonal. Crystals are usually simple rhombohedral, nearly cubic in shape. Penetration twins are fairly common.

OCCURRENCES: Chabazite, along with other zeolites, is found in cavities in volcanic lavas, particularly in basalts. Small glassy crystals have been found at Berufjord, Iceland. Nice large crystals of 1″ to 2″ in size are found at Wasson's Bluff, Nova Scotia, Canada. In U.S., fine crystals from New Jersey. Several localities in Oregon, such as Springfield, Vernonia, and Goble are recorded. Also from Aussig, Czechoslovakia. Good $\frac{1}{4}$″ crystals from basalt are found near Melbourne, Victoria, and from Tasmania, Australia. The Nova Scotia specimens and the New Jersey specimens, particularly near Paterson, are sought. Large crystals have been found at Kilmacolm, Renfrew, Scotland.

Chabazite

STREAK

HARDNESS	white	red	yellow	green	blue	brown	grey	black
1.5–2.0	16, 20, 106	14, 42, 134	44				34	
2.0–2.5	130, 138, 174, 190	28, 48, 50, 134	18, 158	128			34	
2.5–3.0	80, 102, 114, 136, 140, 174, 190		108, 158				26	52
3.0–3.5	80, 96, 102, 104	142		58				
3.5–4.0	56, 86, 88, 92, 94, 118, 122, 132	60	22	100	98	78		32, 46
4.0–4.5	56, 88, 90, 218, 220	60				78		24, 30, 46
4.5–5.0	112, 116, 160, 182, 186, 218, 220							30